SpringerBriefs in Computer Science

SpringerBriefs present concise summaries of cutting-edge research and practical applications across a wide spectrum of fields. Featuring compact volumes of 50 to 125 pages, the series covers a range of content from professional to academic.

Typical topics might include:

- A timely report of state-of-the art analytical techniques
- A bridge between new research results, as published in journal articles, and a contextual literature review
- A snapshot of a hot or emerging topic
- An in-depth case study or clinical example
- A presentation of core concepts that students must understand in order to make independent contributions

Briefs allow authors to present their ideas and readers to absorb them with minimal time investment. Briefs will be published as part of Springer's eBook collection, with millions of users worldwide. In addition, Briefs will be available for individual print and electronic purchase. Briefs are characterized by fast, global electronic dissemination, standard publishing contracts, easy-to-use manuscript preparation and formatting guidelines, and expedited production schedules. We aim for publication 8–12 weeks after acceptance. Both solicited and unsolicited manuscripts are considered for publication in this series.

Indexing: This series is indexed in Scopus, Ei-Compendex, and zbMATH

Yair Neuman

AI for Understanding Context

 Springer

Yair Neuman
Department of Cognitive and Brain Sciences
Ben Gurion University of the Negev
Be'er Sheva, Israel

ISSN 2191-5768   ISSN 2191-5776 (electronic)
SpringerBriefs in Computer Science
ISBN 978-3-031-64209-8   ISBN 978-3-031-64210-4 (eBook)
https://doi.org/10.1007/978-3-031-64210-4

This Springer imprint is published by the registered company Springer Nature Switzerland AG
The registered company address is: Gewerbestrasse 11, 6330 Cham, Switzerland

If disposing of this product, please recycle the paper.

Preface

Context is necessary for understanding human behavior. The etymology of the word context shows us that it emerged from the Latin word *contextus* "a joining together." Without understanding how pieces of information join together, it is impossible to understand human behavior. The current book aims to explain how to use AI to understand context better. It aims at a wide readership, from computer and data scientists to those involved in digital humanities and computational psychology. The interested reader should not, therefore, be surprised if some parts of the text seem trivial, while others beyond their scope of expertise are more challenging.

The first five chapters (1–5) present the theoretical foundations. Chapters 6–8 discuss three case studies in which I elaborate on the ideas and show how to prompt ChatGPT to identify the relevant contextual dimensions and use them better to understand human behavior in all its challenging complexity.

As the book aims to address a broad audience, it combines a scientific style with lively examples and case studies. My case studies are presented as worked examples. Many years ago, I learned from John Sweller's research that there are significant benefits to learning through worked examples. In contrast with some of my academic publications, where I address a scientifically oriented audience to communicate new technologies, here, my focus is on teaching and learning. Therefore, focusing on case studies and worked examples seems fully justified. The reader is invited to a fascinating journey in which I hope to show how AI can help us in "joining together."

Be'er Sheva, Israel   Yair Neuman

Acknowledgments My understanding of context and the complexity of human behavior might have looked different if I had not met Zvi Bekerman as a young student. A quarter of a century later, his teachings still echo in my research. I also thank Yochai Cohen, the software engineer in my DARPA project, for materializing my algorithms in silico. Finally, I thank Springer's editor, Alexandru Ciolan, for his friendly approach to bringing this manuscript into print.

About This Book

Context is necessary for understanding human behavior. However, so far, the concept of context has mostly been treated in a way that lacks any clear relevance for using, developing, and engineering intelligent systems. In this short manuscript, Prof. Neuman explains the importance of context for understanding human behavior, presents a theory of context, and shows how AI, specifically Large Language Models such as GPT, can support our understanding of context when analyzing human behavior as expressed in texts ranging from conversations to short stories. Drawing on years of R&D and academic publications in top-rated journals, Prof. Neuman provides the reader with a deep understanding of context and its modeling for specific challenges, from identifying social norm violations to understanding conversations going awry and stories by great authors. The book may interest a wide variety of readers seeking to incorporate AI into their understanding of human behavior.

Contents

About the Author

Yair Neuman (b. 1968) is a polymath drawing on diverse disciplines to address real-world and academic challenges. He published numerous papers and academic books and was a visiting Professor at M.I.T., University of Toronto, University of Oxford, and Weizmann Institute of Science. Beyond his academic work, he developed state-of-the-art social and cognitive computing algorithms, such as those he developed for IARPA and DARPA.

Chapter 1
The Importance of Context

Abstract We cannot understand human behavior without taking context into account. However, too many projects miss this point either from naïve ignorance, or because they apply an engineering-biased approach, or for the sake of simplicity. In this chapter, I explain why context is important and relevant for computational projects, such as the automatic identification of social norms and their violations and the automatic analysis of human personality.

1.1 Context Is Necessary for Understanding Behavior

It is impossible to understand human behavior without considering the contextual/ situational[1] dimensions in which it occurs. This argument can be illustrated through two simple examples presented long ago.

The first example was given in the 1920s by one of the most important language theorists, Valentine Voloshinov. Voloshinov [1] describes a couple sitting in silence in a room when one suddenly says, "Well!", while the other does not reply. For us, as outsiders, the meaning of this act of communication, the utterance "Well!", is incomprehensible. We learn from this lack of understanding that the meaning of "Well!" is not contained in the word itself. Like a particle residing in a physical space, the meaning of a word emerges only through the semantic field in which it is located.

What do we need to understand the meaning of "Well!"? What we need, suggests Voloshinov, is the non-verbal context. This includes three components:

1. the spatial purview the speakers share,
2. the couple's "common knowledge and understanding of the circumstances," and
3. the couple's "common evaluation" of the circumstances (ibid., p. 11).

Voloshinov provides these contextual dimensions to show how the single word "Well!" may be loaded with meaning. He explains that the couple are glancing out

---

[1]Here, I use the terms "context" and "situation" interchangeably.

Y. Neuman, *AI for Understanding Context*, SpringerBriefs in Computer Science, https://doi.org/10.1007/978-3-031-64210-4_1

the window and see snow falling, and yet it is May. They are both sick of the winter and anxious it will end; therefore, the one word "Well!" expresses a mental state of anticipation they both share and understand.

The example is appealing in its simplicity and illustrates an important point: we must consider contextual/situational dimensions to understand human verbal or non-verbal behavior. Knowing nothing about the situational dimensions, we cannot understand the meaning of "Well!". Without these contextual dimensions, even this relatively simple act of communication is meaningless. The situation (or context) describes the circumstances in which the utterance occurs. Therefore, any act of communication is contextual in that it cannot be understood without considering certain dimensions outside the behavior which is the focus of our attention.

Here is a second example given by the anthropologist and polymath Gregory Bateson in a book published during the 1970s [2]. Bateson describes a mother who habitually rewards her child with ice cream when he eats his spinach. At that time, spinach was not considered an attractive food for children, despite its nutritional benefits. If you recall the animation heroes of that time, *Popeye the Sailor Man* should immediately come to mind. Popeye gained strength by opening a can of spinach and swallowing it in one gulp. Swallowing the spinach, he immediately became superhuman, saving his loved one and beating his ultimate opponent. The mother in Bateson's story does not have Popeye to help her convince her son to eat spinach. Therefore, she rewards the behavior of eating spinach by doling out ice cream.

Here, Bateson is challenging us with an interesting question. What additional information, asks Bateson, do we need to be able to predict whether in the future the child will:

1. come to love or hate spinach,
2. love or hate ice cream, or
3. love or hate Mother?

Notice that the simple situation of a mother rewarding her son for eating spinach may lead to *diametrically opposing* behaviors. For example, the son may come to *love* or *hate* spinach. He may associate the spinach with the caring mother and the rewarding ice cream and, therefore, may come to love the taste of spinach. In contrast, he may associate spinach with its unpleasant and bitter taste and with a mother forcing him to eat something he despises. Therefore, he may hate spinach (and his mother …).

It is important to realize that these two *opposing* behaviors (i.e., love or hate) are both reasonable trajectories for the *same* behavior, the behavior which is the focus of our attention. At the micro level of analysis, this same behavior can potentially lead in totally different directions. In a profound sense, the entropy or the uncertainty associated with this particular behavior is maximal. The outcomes seem theoretically similar to the outcomes of tossing an unbiased coin.

We must focus on additional aspects of the mother–son interaction to predict the child's behavior and reduce the entropy. Bateson shows us, similarly to Voloshinov, that there is not enough information in the interaction itself to remove the uncertainty concerning the child's future behavior. At the micro level of analysis, the uncertainty associated with a piece of behavior, a word, a sentence, etc., is maximal. The behavior which is the focus of our attention is like an isolated particle that has no meaning outside the field in which it exists.

So, what information do we need to remove some of the uncertainty and reduce the entropy? The answer given by Bateson is "context," defined as "the collective term for all those events which tell the organism among what set of *alternatives* he must make his next step" [3, p. 289, my emphasis].

Context is, therefore, associated with a *set of "events"* supporting some *choice* between *alternatives*. Therefore, the concept of context is irrelevant in a fully deterministic world. This concept is relevant for understanding behavior in a non-deterministic world where the same behavior may lead to different trajectories.

While Bateson's definition of context is too general, his idea of context can be understood intuitively. For example, the rewarding mother may be conceived by the child as a control freak forcing him to eat something that he dislikes. In contrast, the child may conceive of his mother as a caring person who gives him the healthy food necessary for his growth. These two "contexts," corresponding to the way the child *represents* the situation in his mind, may lead to different future choices or behaviors (e.g., liking versus disliking spinach).

In sum, any behavior in the focus of our attention cannot be understood without taking some additional information into account. This information, which we call context, aims to *constrain* the potential trajectories (i.e., meaning) resulting from the same behavior and to improve our understanding and prediction. According to Bateson's ideas, understanding context means identifying the alternatives facing the organism and also the supporting information needed to understand how specific alternatives are more likely to follow.

Both examples, given by people from different academic traditions and cultural backgrounds, converge to the same conclusion: to understand human behavior, we must consider the context. The importance of this understanding may be further illustrated in two cases: understanding social norms and their violations and understanding human personality. These two cases are associated with current computational challenges.

The challenge of understanding social norms and their violations underlies DARPA's project of Computational Cultural Understanding.[2] The project is explained as follows:

"The U.S. Government and its agencies [...] operate globally and are in constant contact with diverse cultures. Communicative understanding, not simply of local languages but also of social customs and cultural backgrounds, lies at the heart of Civil Affairs and Military Information Support Operations activities, which together comprise the vast majority of U.S. counterinsurgency and stabilization efforts. Cross-cultural miscommunication not only derails negotiations, but also can be a

---

[2]https://www.darpa.mil/program/computational-cultural-understanding.

contributing factor leading to war. The likelihood of communicative failure increases dramatically where significant social, cultural, or ideological differences exist."

This description emphasizes the importance of understanding other cultures to avoid communication failures that may lead to detrimental results. In this context, the importance of developing technologies for understanding social norms and their violations is clear. However, social norms and their violations are not easy to identify, because they are contextual. For example, walking naked on the beach is illegitimate unless it is a nudist beach, one is an infant, and so on. To understand whether a particular behavior expresses the violation of a social norm, we must take context into account. Given the enormous number of social norms, it is difficult to imagine how we could automatically identify social norm violations. Context may be a key to addressing this challenge.

The second case concerns human personality. Understanding others, on the individual level of analysis, is an important challenge for human beings. This challenge has been amplified in a highly populated world. For example, the availability of expert therapists is extremely limited because we cannot provide individual-based therapy to a mass of people. Creating artificial therapists to provide the first step of diagnosis and treatment may address this difficulty. To develop such tools, AI must correctly represent human personality, which is situational/contextual, as will be explained shortly. The following two cases explain the importance of understanding social norms and human personality contextually. First, I will use the case of social norms, as my DARPA-sponsored project focuses on this subject.

1.2   Hugging, Social Norms, and Dumb Lloyd

The successful American comedy "Dumb and Dumber" (1994)[3] presents the adventures of two dumb characters: Lloyd and Harry. In one of the scenes, Lloyd, played by the hilarious Jim Carrey, is a limousine driver bringing a rich client named Mary Swanson to the airport. When saying goodbye, Lloyd behaves strangely, asking Mary for a hug as if they were family members or close friends. The humor here comes from Lloyd's violation of the social norm that we usually only hug individuals who are emotionally affiliated with us.

Social norms are the unwritten social *rules* and *expectations* that govern behavior within a particular society or group. Therefore, norms are social *regularities* that guide our *expectations*. These norms dictate how individuals should behave in various social situations, encompassing aspects such as manners, customs, values, and beliefs. Mary's facial expression indicates that Lloyd violated a social norm. When Lloyd hugs her affectionately, her face shows surprise and embarrassment. To the amused audience, it is clear that Lloyd's behavior is

---

[3]https://en.wikipedia.org/wiki/dumb_and_dumber.

inappropriate and embarrassing, justifying the view that he is a dumb character. Violating a social norm is not appreciated unless presented in a comedy.

This intuitive understanding of Lloyd's embarrassing behavior is deeply grounded in our evolutionary origins. A constructive approach I recommend for understanding human behavior is to examine non-human organisms that share family resemblances with us. Identifying the similarity with non-human organisms may help us to expose the basic common denominator, thereby removing some of the complexity associated with human behavior.

What can we learn about hugging from other mammals? Hugging is a form of greeting and affection expressed by humans [4] and a behavior we share with other primates, as explained by [5, p. 442, my emphasis]: "In several primate species, including humans, embracing predicts the *level of affiliation between subjects*."

As Lloyd and Mary's level of affiliation is low, embracing is not expected, and when this expectation is violated, our response involves surprise and embarrassment. Therefore, context forms expectations, and when these expectations are violated, we experience surprise. In this case, the relevant context is the *level of affiliation* between a limousine driver and his client. This context *constrains* the expected alternatives for their behavior. Remember Bateson? When saying goodbye at the airport, we expect certain kinds of behavior/alternatives with higher probabilities. When the interaction is between a limousine driver and his client, it is a context directing us to believe that asking for a hug is a less probable alternative than cordially and formally saying goodbye.

Hugging and embracing are fundamental human interactions. It is argued that "cues of social support predict the developmental environment" [6, p. 2], meaning that cues of social support, such as hugging, are informative about the social context to which we should adapt. Lloyd's behavior is awkward as it involves *signaling cues that are incoherent with the context and the social norm,* suggesting that embracing communicates emotional states and social intentions. The emotional states of a limousine driver and his client do not usually support embracing or social expressions of intimate support. Our social and relational expectations are, therefore, formed through our *background knowledge* and *inferences*. When our expectations are violated, we are surprised.

So far, the humoristic aspect of Lloyd's behavior is clear. However, how we understand that this behavior is a violation of a social norm is not trivial, specifically if we are seeking to translate our social understanding into algorithms capable of identifying social norms and their violation by human beings (see, e.g., [7]). For example, imagine a soap opera where the scene occurs in a hospital. The surgeon comes out of the operating theater and informs an old lady that her husband has just passed away. The woman bursts into tears and hugs the surgeon. Is it a violation of a social norm? One may say that the level of affiliation between a surgeon and the patient's wife is low, and therefore it is a case of norm violation. However, one would hardly describe this behavior as norm violation. This example illustrates the tremendous difficulty in forming a simple rule-based system for representing context and understanding social norms and their violation in complex, dynamic, and culturally different human interactions. The enormous number of contexts relative to which a piece of behavior may or may not be considered legitimate invites us to make a deeper inquiry into the way context can be conceptualized and represented, specifically for designing and engineering AI-based technologies. To illustrate these points further, let us move on to the next example.

1.3 The Context of Personality

Human personality involves stable patterns of thought, emotion, and behavior [8]. For example, a person described as an introvert is characterized by a preference to orient themselves toward the inside rather than toward the outside world. Such a person may feel more comfortable in quiet, solitary environments where they can engage in introspection and reflection and enjoy less energetically demanding activities. These might include reading, writing, or hobbies that allow for individual exploration and creativity.

When defining personality as a stable pattern, we should ask whether this definition is empirically grounded. In other words, do we really have personalities? The idea that our personality is a stable mental structure *inside* us has been challenged by Walter Mischel, the famous psychologist known for *The Marshmallow Test*. Mischel describes the *Personality Paradox* as involving the discrepancies between the expressions of consistency expected from personality and those actually found [9]. The idea is simple to understand and explain. Mischel takes us back to the 1920s when Theodore Newcomb [10] studied extroversion and introversion in 51 boys in 21 situations in a summer camp. If there is an extrovert/ introvert personality, then one should expect measurements of the *same* personality in *different* situations to be positively and highly correlated. This expectation, or hypothesis, is naturally derived from the definition of personality. As personality reflects *stable* patterns, it is supposed to go with us wherever we go. Newcomb's finding was remarkable. The average correlation between personality measurements across the situations was close to zero. No consistency was found even for a personality dimension like introversion, which is considered so basic that it even characterizes some non-human organisms.

In the "classic view" of personality, explains Mischel, "the basic qualities of the person are assumed to be *independent* of, and unconnected with, *situations*" (ibid., p. 4, my emphasis). However, there is no personality which is dissociated from a situation. In the late 1960s, Mischel published a book with what was considered a heretical thesis: personality is contextual [11]. To further explain this idea, let me use a personal anecdote.

Years ago, I hosted a famous psychologist who arrived from abroad to give a seminar at my university. I arrived at the airport to pick him up and met a shy, quiet, and highly introverted person. We hosted my colleague for a couple of days, and he behaved throughout as a shy introvert. However, when he gave his seminar, something miraculous happened, and the ultimate introvert magically transformed before my eyes. He became highly extroverted and energetic with different behavior, intonation, and body language. I saw a clear expression of Mischel's contextual theory of personality. The psychologist felt insecure in personal social interactions and behaved like an introvert. However, when presenting to an academic audience, he could express his natural talent, passion, and intellectual strength. He therefore felt secure and expressed an extrovert type of personality.

The contextual approach to personality—the idea that there may be situations in which personality may vary—encountered bitter opposition, and allegedly for a good reason: saying that personality is contextual seems to imply that personality does not exist. For comparison, think about a mathematician struggling to establish the scientific foundations of mathematics and suddenly being told that the meaning of "1" is not absolute but somehow context dependent. However, Mischel anticipated this criticism and suggested that stability can be identified *within personality–context* relationships, where the term "context" describes an abstract and general pattern that explains the *variability* and *invariability* of personality. Returning to the previous anecdote, the psychologist's introverted personality was expressed in the context of INTERPERSONAL RELATIONSHIPS. The *invariability* of this personality could have been identified in this repeating context. However, *personality variability* was expressed in INTERPERSONAL RELATIONSHIPS versus ACADEMIC CONTEXT. My understanding of my colleague's personality was enriched by identifying its contextual/situational aspects: stability in a specific context and variability across different contexts.

The number of potential contexts we need to understand a given piece of behavior might be enormous. In practice, and when examined from a higher level of abstraction, we can identify a finite and limited number of contexts that may significantly enrich our understanding of a specific form of behavior. In other words, identifying a few contexts may significantly reduce the entropy and improve our understanding. Unfortunately, the majority of personality research does not follow Mischel's approach. It is easier to believe in simple foundations than to empirically address a complex reality.

Now, imagine an automatic system that analyzes the behavior of a couple. For example, imagine a system analyzing the couple's behavior in *Scenes from a Marriage*[4] In this drama, we follow the lives of a couple, Jonathan and Mira. At the beginning, we see a happy couple composed of individuals with well-defined and stable personalities. However, various things happen along the way. For instance, there are contexts where Jonathan, the self-assured professor of philosophy, turns into a needy baby. In what contexts do we observe Jonathan's different personality dimensions? Automatically identifying these contexts is no trivial task, and developing a system that can elucidate Mischel's personality-in-context is currently an open challenge. However, it is an important challenge, specifically in the context of computational personality analysis (see, e.g., [12]). An automatic identification of human personality might well be uninformative unless identified as part of the personality–context association.

---

[4]https://en.wikipedia.org/wiki/Scenes_from_a_Marriage_(American_miniseries).

This opening chapter has explained the importance of contextual understanding and the enormous challenges accompanying the modeling of contextual understanding when designing intelligent systems. The next chapter will delve deeper into context and integrate past ideas and theories to understand better what is involved in context.

References

1. Voloshinov, V.: Discourse in life and discourse in poetry. In: Shukman, A. (ed.) Bakhtin School Paper. RPT Publications, Oxford (1983/1926)
2. Bateson, G.: Steps to an Ecology of Mind. Granada, London (1973)
3. Bateson, G.: Steps to an Ecology of Mind: Collected Essays in Anthropology, Psychiatry, Evolution, and Epistemology. University of Chicago Press, Chicago (2000)
4. Forsell, L.M., Åström, J.A.: Meanings of hugging: from greeting behavior to touching implications. Compr Psychol 1, 02–17 (2012)
5. Pallante, V., Ferrari, P.F., Gamba, M., Palagi, E.: Embracing in a female-bonded monkey species (*Theropithecus gelada*). J. Comp. Psychol. 133(4), 442 (2019)
6. Snell-Rood, E., Snell-Rood, C.: The developmental support hypothesis: adaptive plasticity in neural development in response to cues of social support. Philos. Trans. R. Soc. B 375(1803), 20190491 (2020)
7. Neuman, Y., Cohen, Y.: AI for identifying social norm violations. Sci. Rep. 13(1), 8103 (2023)
8. Funder, D.C.: The Personality Puzzle. WW Norton & Co., New York (1997)
9. Mischel, W.: Toward an integrative science of the person. Annu. Rev. Psychol. 55, 1–22 (2004)
10. Newcombe, T.M.: Consistency of Certain Extrovert–Introvert Behavior Patterns in 51 Problem Boys. Columbia University, Teachers College, Bur. Publ., New York (1929)
11. Mischel, W.: Personality and Assessment. Wiley, New York (1968)
12. Neuman, Y.: Computational Personality Analysis. Springer, New York (2016)

Chapter 2
Toward a Theory of Context

Abstract This chapter presents the meaning of context and explains how to model it. We can understand context as a set of relations. These relations function as constraints reducing the entropy of alternatives by priming specific alternatives over others. It also explains how much context we need to address context-based computational challenges.

2.1 Why Do We Need Context?

The difficulty in understanding human behavior, specifically concerning texts, either written or spoken, involves foundational questions of understanding and meaning. To avoid lengthy philosophical discussions which have no place here, I start with a "naïve" question: Why do we need context? I adopt a general scientific perspective and apply it to the particular context of human behavior.

In communicating with others, human behavior, whether physical, verbal, or mental, is built upon a *general* platform of communication aimed to serve communication in a *particular* situation. When meeting a friend and saying, "Hello, Saadia, good to see you," I exploit resources we share, from the *mental lexicon* to the *grammatical structure* used to organize the words into a sequence.

Communication is possible only through shared knowledge and general conceptual patterns. For example, the utterance "Hello" is a specific type of *speech act* called a GREETING. Understanding the greeting means that we share the same lexicon with the speaker. It also means we can represent the utterance as a particular instance of GREETING. Here is another example. When Lloyd hugs Mary, he does not invent a new style of communication but uses the act of EMBRACING, signaling a high level of SOCIAL AFFILIATION. Therefore, communication is always grounded on a general platform, and for a good reason: we need a general and shared platform to communicate with *others* across *contexts*. However, a *general* platform like natural language involves built-in flexibility to support communication in *particular* situations, because communication always happens here and now.

Think about the number of words we use to communicate with each other. For instance, the sign "I" is the first-person pronoun I use to communicate my first-person perspective of the world. However, the person to whom this "I" refers has changed significantly throughout his life. I am not the same person I was 50 years ago, 30 years ago, or ten years ago. As beautifully explained by [1, p. 28]:

"Much as Peter Pan's shadow is sewn to his body, the "I" is the needle that stitches the abstraction of language to the particularity of the lived experience."

This insightful idea by Holquist grasps the essence required to understand the in-built tension between the "abstraction of language" and the "particularity of the lived experience." As I shall explain later, context aims to resolve this tension.

Holquist's insight applies to every word we use. Each word is a device that stitches the abstraction of language to the particularity of the specific situation in which it is used. An inevitable question is why not use particular words for particular experiences. Why use the word "sky" to signify an entity that changes with the seasons instead of inventing different words for each particular sky configuration? The answer is simple: you cannot store the enormous number of words you would need in your memory to describe each particular entity. This is because the accompanying *cognitive load* contradicts the principle of cognitive economy, which aims to reduce our energy expenditure. Here, we face a conflict between the general/abstract and the particular. On the one hand, we need a general and abstract platform that may be used to communicate with others and with minimal cognitive effort. On the other hand, communication is always about the particularity of the here and now; therefore, the abstract and general language must be stitched to the particularity of the situation.

The above-mentioned conflict is expressed in the notion of *polysemy,* which is the *coexistence* of several possible meanings/functions for the same word or phrase. Interestingly, polysemy exists in biology, too [2], where the same biological entity has different functions expressed in different contexts. Life is expensive, and one should strive to save energy whenever possible. Moreover, polysemy allows any living system to gain flexibility in response, and this is important for effectively fine-tuning behavior to a particular context. The same word can have different senses in different contexts (e.g., the palm of your hand vs. the tree called a palm). This "logic" of communication explains why the behavior of all living systems is heavily grounded in context [2]. The representation of a context aims to balance the general and multi-functional aspects of the communication platform with the particular needs of the here and now. This is why a contextual representation is built into living systems, from the immune system [2] to the human personality. In sum, context is required to stitch the abstractness of language to the particularity of the "lived experience" or the actual situation in which the communication takes place.

The act of communication is the *focus* of our attention. The context is the set of signals that exists in the *background*. These signals should be few and abstract to avoid the number of contexts becoming too great. In other words, when representing situations, we should be able to do this with an *optimal level of abstraction* in a way that is informative enough to generalize beyond particular contexts and, at the same time, use a representation that is not too general to avoid abstract nonsense in which important differences are lost. This challenges those dealing with R&D in human behavior and intelligent systems. But the challenge is broader: How can situations be analyzed to support the understanding of human behavior and the development of technologies centered around human or human-like behaviors? From the design of robots to assist the elderly to the design of digital personas for the gaming industry, computational situational analysis is an important aspect of current and future technological innovation.

In sum, we need context because communication happens here and now, and communicating through general and abstract platforms requires us to ground these general tools in a specific situation. *The context is the set of signals mediating between the general and the particular.* The next step is, therefore, to understand how to model context.

2.2 Modeling Context

I start with the idea, presented by the famous John von Neuman, that science is about modeling [3, p. 683]:

The sciences do not try to explain, they hardly even try to interpret, they mainly make models.

Scientific models are relational structures in which a small number of concepts and their well-defined relations (e.g., the idea that bodies are attracted as a function of their mass) form a simple representation of an "observed phenomena" (e.g., gravity). Different scientific fields produce different types of scientific models. Those produced in physics are different from those produced in biology. However, there is a general prototype of a model. First, it is a *simple conceptual structure* involving a few *related* elements. Second, the elements and their relationships are *well-defined and empirically measurable*. Third, *prediction* and *control* are the ultimate expression of a good model in the empirical sciences. If you can predict the trajectory of a clown shot out of a circus cannon, then your model works. If you can control the phenotype of an organism by editing one of its genes using a good model of the gene's expression, then your model works. A model is not just a nice picture of the world. It is a map which is supposed to direct you to your destination.

From a cognitive perspective, human beings, like scientists, may be conceived as modelers. They build relatively simple representations of observed phenomena and expect them to work. What does it mean "to work" in the context of human representations? At the minimal level, a model should somehow make sense, whether in a scientific or a non-scientific way. Astrology gives people a non-scientific model of human beings, but it has the advantage of order over chaos; it makes sense of the world. In other words, like a good story, a model is expected to organize things into a structured and *coherent* pattern. The model shows us how things are related and form something we can identify as a structure. Even if it is invalid, a good story is psychologically better than chaos.

The second aspect of a cognitive model is prediction. Modeling human beings by identifying their personality type, we may predict, for instance, that it is more likely for an introvert to favor jazz music over rap. A good model is tested by reducing the world's entropy (i.e., uncertainty). This reduction in uncertainty is expressed in prediction.

The third criterion for a successful cognitive model is control. For example, when conversing, we repeatedly perform minor and major adjustments [4], similar to those we perform to maintain our balance or to those we apply when dancing with our partner. Our models use feedback loops to control our verbal and non-verbal moves. In a study I performed during my DARPA project [5], I showed that a change in emotion during a conversation can be successfully predicted by using a simple model explaining how interlocutors apply feedback loops to adjust their emotions to the emotion of the other. A good model, whether in science or human behavior, is also judged by the ability to control. If you have a good model of yourself and your interlocutors in a conversation, you can control the conversation better.

Drawing on this general scientific perspective, we may say that to understand means to be able to model by organizing some elements into a structure/pattern and using the model to predict and control. The extent to which we understand something depends on our ability to form a correct representation, such as a scientific model, to derive valid predictions from the model, and to control the phenomenon. Let me illustrate these general scientific ideas in the context of human verbal interactions.

Sarah tells her husband, Abraham: "I was offended by your behavior." Abraham should form a model to understand his wife. First, he should transform the linear sequence of words (i.e., the token) into a structure involving the relationships between the words. The French linguist Tesnière argued [6, p. xii, emphasis mine] that "understanding language involves transforming *linear* order to *structural* order." This is precisely the pattern formation phase we discussed previously. The elements of the observed phenomenon that appear in a linear order are transformed and represented as a *structure*. In the context of linguistics, Tesnière proposed a relational structure in which the verb plays the most important role. Using ChatGPT to analyze the above-mentioned utterance according to Tesnière's theory, we get:

Verb with highest valence: offended

Subjects: I

Objects: your behavior

The representation of the sequence as a structure reveals that the utterance is about the action of offending and the speaker being offended by the addressee's behavior. If Abraham correctly represents the structure, he should understand that his wife's utterance is centered around the activity of offending her. So? What is next? To understand what is next, we should go beyond the structure.

Both interlocutors, Sarah and Abraham, should know that language is functional because an utterance is an *action* performed through language. In other words, they should both understand that, in the context of social interactions, people do not produce utterances for nothing. This is the *pragmatic* aspect of language. When producing an utterance, people do something. They act, and when they act, they expect a specific response. The term describing this action is *"speech act."* A speech act is a general action performed through speech or communication. For example, if I ask, "Can you please bring me my glasses?" I am performing the speech act of REQUESTING. When I ask, "How long should it take for the delivery?" I am performing the speech act of asking questions and anticipating an answer that provides information.

The speech act expressed by the utterance "I was offended by your behavior" may be described as an *assertion* or *statement*. In this utterance, Sarah expresses a personal feeling (being offended) caused by her husband's behavior (your behavior). However, the same utterance could also be interpreted as a *complaint* or *criticism*, depending on the context and tone of the communication. The same utterance can be classified into different speech acts inviting different responses. Therefore, Abraham represents the utterance by "choosing" one of the alternatives: assertion, criticism, or complaint. Understanding his wife, Abraham should first represent his wife's utterance by understanding the centrality of the verb she used and by abstracting the correct speech act communicated by her utterance. Correctly identifying the speech act depends on additional factors, such as understanding Sarah's intention; without these additional factors, Abraham cannot understand Sarah and effectively respond to her utterance. Now, we better understand the pragmatics of language, which focuses on how meaning is produced in context [7].

Understanding the utterance as a speech act expressing a complaint, what is expected from Abraham? It depends … on context, of course. Under certain constraints, Abraham may choose to respond by apologizing. APOLOGY is a speech act where the speaker expresses regret or remorse, or acknowledges responsibility for a past action or event that has caused harm, inconvenience, or offense to another person. There are, however, several alternatives for responding to a complaint. Possible speech acts expected in response to a complaint include:

1. **Acknowledgment**: The listener acknowledges the complaint by responding with phrases such as "I understand," "I hear you," or "I see what you are saying."
2. **Empathy**: The listener shows empathy by understanding the speaker's feelings or situation, for example, by saying, "That sounds frustrating," or "I'm sorry to hear that."
3. **Apology**: An apology may be appropriate if the complaint is valid and the listener is responsible. The listener may say, "I apologize for the inconvenience," or "I'm sorry that happened."
4. **Explanation or clarification**: The listener may explain or clarify if the complaint is due to a misunderstanding or lack of information. This could involve

saying, "Let me explain why that happened," or "Here's some additional information that might help."

5. **Resolution or action**: Depending on the nature of the complaint, the listener may offer a solution or promise to take action to address the issue. They might say, "Let's see how we can fix this," or "I'll make sure to address that right away."

6. **Further discussion**: Sometimes, the listener may suggest further discussion to understand the complaint better or explore potential solutions. They might say, "Can we talk more about this?" or "Would you like to discuss possible solutions?"

Predicting Abraham's response shifts the modeling process to Sarah or even to us as outside observers. Our prediction depends on factors external to the utterance and its structural representation. If we know something about these factors (i.e., context), we can improve our prediction of Abraham's response and even control it. For example, knowing that Abraham is sensitive and caring, Sarah may say: "Honey, your behavior offended me." In this context, she emphasizes her love and the fact that she has been offended by someone she loves. This context invites an apology. Alternatively, we may learn that Sarah and Abraham are continuously involved in power struggles. In this context, Sarah can emphasize her criticism by saying: "You insensitive and vicious bastard, your behavior offended me!" This insulting utterance may lead to a different response. Different utterances may lead to different responses, depending on the context, which *primes* specific responses over others.

Let me sum up. Understanding the utterance involves a modeling process. This process first transforms the utterance into a relational structure. Representation is the first phase of modeling. Next, we have a process of abstraction where the utterance is represented in Abraham's mind as one of several possible speech acts. Given this representation, he may use several possible speech acts in return. Predicting which speech acts will be used to represent the utterance and respond to it depends on additional factors we call "context." More generally, modeling context requires us to represent the given behavior as a set of relations and use additional factors to predict a response.

2.3 Becker on Context

The insightful anthropological linguist Alton Becker [8] put forward the idea that context is a *set of relations* that minimally involves four dimensions. The first dimension is *coherence*. It is the set of relations between the units of the text (e.g., words, sentences, etc.). This is the same as Tesnière's idea of a structure. The relation between the units of the text is the *coherence* of the text and the way its elements form a pattern. If Sarah says: "Honey, you offended me. I expect an apology," then the relation between the first part of the utterance and the second part is coherent and clarifies her intention, which is a request for an apology. The utterance is coherent because the second part is "naturally" derived from the first part. In other words, the utterance, as a whole, makes sense. It makes sense because, in our mind, we can represent it as a general pattern of requesting an apology. Coherence involves inferences that we may draw from the text. It is reasonable to guess that when Sarah says: "Honey, you offended me," she expects an apology. In terms of textual entailment, the hypothesis "I expect an apology" is entailed with a high probability from the premise "Honey, you offended me."

According to Becker, the second dimension of context involves the relation of the text to other texts. To understand this dimension, we must first emphasize that communication is always *particular*; it happens here and now in a specific setting that can never be replicated. Metaphorically, we can say that in actual communication, we "speak the present." However, as explained in [9, p. 140, emphasis mine], "one can never wholly speak the present. Even everyday language is highly conventional, far more *constrained* than we usually think." In other words, the past constrains our concrete and present interactions. We always "speak the past." What does it mean to say that we speak the past? It simply means that when we are involved in actual and particular communication, we use general and abstract patterns acquired in the past. This is the *common knowledge* discussed by Voloshinov as one of the dimensions of context.

The relation of the text to other texts can be understood as the constraints operated by past patterns on the particular and present act of communication. When dealing with LLMs, we understand this relation between present and past texts perfectly. When I ask ChatGPT to complete the utterance: "In Italy, they usually serve wine with cheese and …" the model can guess that a reasonable word is "prosciutto." This guess is one of several possible alternatives. Although we do not know the exact word to fill the slot, we can guess some words with higher probability. Other words are excluded from the guess, given their low probability. The ChatGPT model holds in its mind the representation of previous texts. In this context, much less likely would be: "In Italy, they usually serve wine with cheese and PEPSI." Guessing the missing or the masked word is possible using a model built on previous texts. Without memory, we are doomed to live every second anew, and memory means that some older texts, whether abstract or concrete, are used to help us create and understand the meaning of a present text. The previous texts can be concrete, such as proverbs kept in memory, or abstract patterns, such as speech acts that involve a higher level of abstraction.

The third dimension is the "intent of the text builder," which may be interpreted as the action the text builder aims to produce. Using the example of Sarah and Abraham, we should understand what Sarah is expecting as a response to her utterance. Is she expecting an apology? Is she expecting anger to get even with Abraham? In a later conceptualization of context, I will describe this dimension as *expectation*.

According to Becker, the fourth and final dimension is *reference*, which describes the text's relation to other events. Voloshinov describes the spatial purview as one of the context dimensions, the same as the referential dimension of context. Later, I will describe this dimension, which concerns the location, the objects present, and the ambiance.

How do these dimensions contribute to understanding the meaning of an utterance? Understanding a text means using a model that "works" (see [5]). Therefore, understanding is judged according to the pragmatic implications of language use. Becker emphasizes this point when he discusses the pathologies of language. Language, argues Becker, operates under the constraints of the context dimensions previously presented. However, "in madness," where understanding fails, "these constraints are misunderstood."

Let us assume that Sarah expected Abraham to apologize, but Abraham failed to understand the *meaning* of her utterance, which is described here as the *expected response*. His misunderstanding is expressed by his failure to understand the constraint of intention and Sarah's expectations. This misunderstanding can be "pathologically" expressed by responding with retaliating criticism rather than by apologizing. His response indicates the meaning he mistakenly attributed to Sarah's intention. In other words, a failure to understand may be expressed in communication breakdowns with all their practical implications.

Given the arguments given so far, we can define context through the following points:

- *A set of relations*
- *functioning as constraints*
- *reducing the entropy of alternatives*
- *by priming specific alternatives over others*

This definition is broad enough to allow us to proceed but specific enough to guide us in modeling a context. Whenever human behavior involves several alternatives, these alternatives constitute an uncertainty that must be reduced. The context is thus a set of relations that functionally helps us reduce the uncertainty or entropy associated with these alternatives.

2.4   Context and Meaning

So far, I have explained the importance of context for understanding human behavior. Here, I would like to delve deeper into the meaning of context by adopting a *pragmatic* approach to meaning. To begin with, let us return to Gregory Bateson. Bateson was one of the first to realize that, in the realm of the living, responses are brought by information [10] and that information is meaningful only by being *consequential*. Holquist explained the same idea: "Nothing means anything until it achieves a response" [1, p. 48]. This is a critical point epitomizing the pragmatic approach to meaning. The meaning of a specific piece of behavior is consequential in that it can be elucidated only through the response it generates. Meaning is, therefore, contextual, social-interactional, and consequential. For example, when Lloyd hugs Mary, she responds with embarrassment. Mary's response (i.e., embarrassment) *loads* Lloyd's behavior/action with meaning. In itself, Lloyd's behavior is devoid of meaning. The behavior becomes meaningful only through interacting with Mary and giving rise to her specific response. Her response signals that this behavior is inappropriate. The response could have been different. Instead of embarrassment, Mary could have responded with approval or with indifference. However, the context of two strangers in the airport largely determines and constrains her specific response. The context of two strangers vs. two close individuals is a difference that makes a difference. Context, though, is not deterministic. For instance, Mary may be an exceptionally open-minded individual who responds to Lloyd's hugging with approval and amusement rather than embarrassment. In this context, Mary's response may be an *exception*, telling us something interesting about Mary. We can infer, for example, that Mary's response indicates an open-minded personality.

Opposition theory suggests that binary oppositions (e.g., love vs. hate) underlie basic cognitive and linguistic processes [11]. Recall the example of the child and the ice cream, where two opposing behaviors (i.e., love vs. hate) could have followed the same mother–child interaction. Adopting the perspective of opposition theory (see, e.g., [11]), Mary's anticipated response (i.e., alternatives) could have been either *dismissive* or *acceptive*. The expected response to a hug by a stranger is dismissive, because the social norm is that hugging is expected between affiliated individuals. If Mary had responded to the hug with an accepting response, it would have been considered an example of *markedness*.

As Wikipedia explains,[1] "In linguistics and social sciences, *markedness*is the state of standing out as nontypical or divergent as opposed to regular or common." Shannon's classical definition of information is grounded in surprise. We may suggest that markedness or informativeness is evident in a given context whenever a specific observed behavior is *entailed* with a lower level of probability in comparison with its opposite (e.g., embarrassment vs. approval) or other less probable alternatives. Given the context of two strangers, the more likely response to a hug is embarrassment. If the other and less probable response is observed, then it is a surprising behavior telling us something interesting and informative.

At this point, we may identify four important points. First, given a specific piece of behavior, we generate hypotheses about the individuals involved in the interaction and their relationships. Representing Lloyd's behavior using a predicate–argument structure, we can write:

$$HUGS\ (LLOYD, MARY)$$

And abstracting the arguments, we can represent the structure as:

$$HUGS\ (PERSON\_A, PERSON\_B)$$

Now, if one person hugs another, what else can we learn about their relationships? We may hypothesize, for example, that:

---

[1]https://en.wikipedia.org/wiki/markedness.

LOVES (PERSON_A, PERSON_B)

Second, we may examine the coherence of this behavior with our previous knowledge, which is another aspect of the context. Two propositions providing contextual knowledge are:

1. Lloyd is a Limousine driver.
2. Mary is Lloyd's client.

Now, we can ask the following questions: Given propositions (1) and (2), how likely is it that Lloyd will hug Mary? I asked ChatGPT this question, and it answered:

To analyze the likelihood of observing the behavior where a limousine driver (L) hugs their client (M), we need to consider several factors based on social norms, professional boundaries, and the specific context of their relationship. Here are key points to consider:

1. **Professional boundaries**: In most professional contexts, especially in service industries like limousine driving, maintaining a certain formality and respecting personal space is expected. Physical contact, such as hugging, is typically outside the norm unless there is a pre-existing relationship or mutual understanding that such behavior is acceptable.
2. **Cultural norms**: The acceptability of physical contact varies widely across different cultures. In some cultures, a hug might be a common way to greet someone, even in a professional setting, while in others, it could be considered inappropriate or too intimate for such a relationship.
3. **Personal relationship**: If L and M have developed a friendship beyond the professional client–service provider relationship or knew each other personally before the professional interaction, a hug might be more likely.
4. **Context of the interaction**: The specific situation could also influence the likelihood of a hug. For example, suppose M is celebrating a significant life event (like a wedding), and L is significantly part of the day. In that case, a congratulatory hug might not be out of place.
5. **Individual personalities and preferences**: Some individuals are more inclined towards physical gestures of friendliness or gratitude than others. A hug could occur if both L and M are such individuals and the gesture is mutually understood to be appropriate.

ChatGPT identified the contextual dimensions required to answer the question. However, it also pointed out that "the default assumption based on professional norms in many societies would suggest that a hug between L and M is *less likely*." Therefore, Lloyd's behavior is an example of markedness, a surprising behavior indicating one crucial aspect of his stupidity: a lack of understanding of social norms.

Third, we may infer Mary's expected response. If propositions (1) and (2) are true, and Lloyd hugs Mary, what is Mary's anticipated response? The expected response, at least in emotional terms, is embarrassment. A social norm has been violated.

Fourth, we may ask whether Mary's response is coherent with the context or involves markedness, telling us something interesting about Mary. Propositions (1) and (2) are the context. The behavior we seek to understand is "Lloyd hugs Mary." The anticipated behavior that loads the behavior with meaning is Mary's response. Indeed, Mary's response is coherent with the context as given for the moment.

Let me sum up what we have learned so far:

1. *Understanding* in the context of human behavior, specifically in human inter-actions, is interpreted in terms of *modeling*.
2. *Modeling* involves the representation of a structure and prediction through inference.
3. Modeling seems to work if it meets the criteria of representation, prediction, and control.
4. *Context* is the set of *constraints* reducing the entropy of *alternatives*.
5. *Meaning* is expressed in the response and involves the proper use of constraints for modeling the interaction.
6. *Failed understanding*, similar to failed modeling, is evident when representation and prediction, "expected to work," fail to deliver the results, for example, in conversations going awry.

2.5 How Much Context Do We Need?

The discussion so far has not addressed the question of how much context we need to model a situation. In one of his most insightful observations, Bateson writes:

"Suppose I am a blind man, and I use a stick. I go tap, tap, tap. Where do I start? Is my mental system bounded at the handle of the stick? Is it bounded by my skin?" [12, p. 434].

Bateson answers these questions by describing them as nonsense. "The way to delineate the system," says Bateson, "is to draw the limiting line in such a way that you do not cut any of these pathways in ways which leave things inexplicable" (ibid.). In other words, use as much context as you need to accomplish your task. Luckily, the world is structured in such a way that our minds have learned to identify a limited amount of information relevant to behavior. The potential number of contexts required for understanding a specific human interaction is in practice finite. When the situation is more complex, we trivially need more contextual information, and this may be difficult or impossible to acquire. However, our social life is built upon *repeating patterns* and the possibility of *correction*. While complete understanding is beyond our ability, a reasonable level of understanding may be within our reach.

In the following chapters, I present several novel directions relevant to com-putational situation analysis. I do not focus on a single direction but present dif-ferent ways in which context can be modeled and used through AI to support our understanding of human behavior. My approach relies heavily on LLMs that revolutionize the field, specifically ChatGPT. Therefore, the next chapter will explain these models and why I consider them to be *contextual machines*.

References

1. Holquist, M.: Dialogism. Routledge, London (1990)
2. Cohen, I.R.: Tending Adam's Garden: Evolving the Cognitive Immune Self. Elsevier, New York (2000)
3. von Neumann, J.: Method in the physical sciences. In: Bródy, F., Vámos, T. (eds.) *The Neumann Compendium*. World Scientific, Singapore (1995)
4. Clark, H.H.: Using Language. Cambridge University Press, Cambridge (1996)
5. Neuman, Y., Cohen, Y.: Predicting change in emotion through ordinal patterns and simple symbolic expressions. Mathematics **10**(13), 2253 (2022)
6. Tesnière, L.: Elements of Structural Syntax. John Benjamins Publishing Company (2015)
7. Mey, J.L.: Pragmatics: An Introduction. Blackwell, Oxford and Malden (2020)
8. Becker, A.L.: Beyond Translation. University of Michigan Press, Michigan (2000)
9. Becker, A.L.: Text-building, epistemology, and aesthetics in Javanese shadow theatre. Dispositio **5**(13/14), 137–168 (1980)
10. Harries-Jones, P.: A Recursive Vision: Ecological Understanding and Gregory Bateson. University of Toronto Press, Toronto (1995)
11. Assaf, D., Cohen, Y., Danesi, M., Neuman, Y.: Opposition theory and computational semiotics. Sign Syst Stud **43**, 159–172 (2015)
12. Bateson, G.: Steps to an Ecology of Mind. Granada, London (1973)

Chapter 3
Large Language Models as Contextual Machines

Abstract Large Language Models, such as ChatGPT, are revolutionizing the field of AI. In this chapter, I explain that LLMs are not just deep neural networks capable of guessing words in a sequence. LLMs are also contextual machines. However, they are contextual machines working at the token level of analysis without directly processing information on the conceptual level of analysis. The promise of using LLMs for a contextual analysis is discussed by emphasizing the importance of expert-based prompting.

3.1 LLMs are Deep Neural Networks

Generations of scholars and researchers have spent time and energy discussing the way texts should be interpreted. LLMs revolutionize and leverage our contextual understanding and our ability to interpret texts, and for a very good reason: they are deep neural networks with a unique architecture (i.e., transformers) designed for contextual representation. In this chapter, I discuss three important aspects of LLMs: first, LLMs are deep neural networks; second, they can be considered as probabilistic guessing machines; and third, they are contextual machines capable of contextual analysis through their self-attention mechanism. For an excellent introduction to machine learning and neural networks, I highly recommend [1]. Deep learning is introduced in several books, from [2] and [3] to more recent texts such as [4]. LLMs are explained by [5] and surveyed by [6] in a highly cited paper. Starmer's StatQuest also includes excellent tutorial videos, such as the one describing transformers that are the foundation of ChatGPT[1] A short introduction to ChatGPT is presented in [7], focusing on a specific aspect of ChatGPT. The interested reader is invited to extend their knowledge using these resources.

Let us assume you want to develop an intelligent system that automatically identifies whether a person is American or British. You are sure it is possible to identify the person by analyzing their pronunciation. For example, you know that American and British people pronounce "tomato" differently:

---

[1]https://www.youtube.com/watch?v=zxQyTK8quyY&list=WL&index=16&t=10s.

American: tuh-may-toh
British: tuh-mah-toh

Therefore, by analyzing the different phonemes, you may have a perfect hint of whether someone is British or American. When developing the system, you build a *model* to predict a categorical variable with discrete values (i.e., the speaker's nationality). The model includes certain measurable variables (i.e., the phonemes in a word) that you use to predict another variable. Your model assumes that there is some kind of rule, deterministic or probabilistic, associating your explanatory independent variable (e.g., the phonemes composing the word) with the dependent variable you wish to predict (e.g., the speaker's nationality). This is a simple example of a classification model. When developing this model, you build a function associating your input to your output (i.e., nationality). However, you do not have to actually identify the associating function. In a supervised machine learning (ML), you provide the machine with labeled examples. Each case includes the nationality of the subject (i.e., American vs. British) and the measurable phonemes. It is left up to the machine to learn the associating rule from the data.

Here is another simple example of a classification model. If you want to predict a person's weight (i.e., the dependent variable), you can use a simple model with two independent variables: age and height. The reason for using this predictive variable is that taller people generally weigh more than shorter people, and older people generally weigh more than younger people (e.g., babies). These two independent variables can be combined in a simple linear model, where the predicted dependent variable is a linear function of the two independent variables. This means that weight equals height multiplied by some coefficient, plus age, also multiplied by a coefficient. The coefficients represent the weight change associated with a one-unit change in height and age, respectively. When training the ML model using this linear function, the machine learns the coefficients that provide us with the best predictive results.

As you may recall from my previous explanation, a model is not just a way of associating some measurable variables. A model must work. It means that we strive to build a model that successfully predicts the dependent variable. Here, I introduce two important ideas: learning and optimization. Like any natural intelligent system, our modern ML models learn to provide the best possible model. They learn by getting some independent variables as input and optimizing the coefficients. Let me explain this process step by step, this time with a new example.

You are a student asked to build an intelligent system that listens to pub conversations and decides whether the speaker is from Scotland or Wales. You record conversations and extract the words spoken. For simplicity, assume that you have only three words:

(1) Slàinte: A toast, which means "health" in Scottish Gaelic, is used when drinking.
(2) Cwtch (Walsh): A hug or cuddle, but with a sense of a safe, protected space.
(3) Beer

These are your independent variables. You record pub conversations where each speaker is labeled "Scotland" or "Wales." You also have the percentage of each word used by each speaker. This is your dataset. Next, you use an ML architecture known as a neural network, and in fact, a neural network that has a very simple architecture. The first layer of the network includes the measurements of your independent variables, such as the percentage of cases in which each word was used in a conversation. At the end of the network's architecture, we have the dependent variable, which, in our case, has two values: Scotland or Wales. In between, we have a layer of "nodes" that are artificial neurons activated by previous layers, as shown in Fig. 3.1.

The magic happens in the *hidden* layer of nodes between the given measurables and the final node delivering the dependent variable. During the learning process, the network is exposed to cases (i.e., people) with their given ethnicity (Scotland or Wales) and the percentage of cases in which they used each word in their pub conversations. Through a process known as back-propagation, the network learns to adjust its weights to optimize its performance.

Our network is simple and includes only three variables and one hidden layer. However, deep neural networks are much more sophisticated. They do not have just a few variables and a single hidden layer. A deep neural network is a very complex function that associates the independent variables with the dependent variable using many hidden layers. This is not a simple function like the one associating weight with age and height. The function computed by a deep neural network is so complex that our limited minds cannot understand it. This is the "black box" of deep neural networks. In the above example, I used three words and one intermediate layer. However, deep neural networks are much more complex. They may digest hundreds or thousands of independent variables through many intermediate

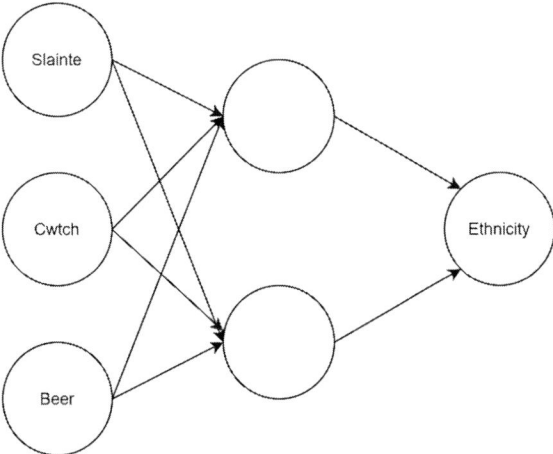

Fig. 3.1 The architecture of our network (image by the author)

and hidden layers of nodes. In sum, our discussion focuses on a specific type of ML model, neural networks, and on a specific type of these, deep neural networks.

The magic of deep neural networks is in the process of *optimization*. The network learns to improve its prediction by fine-tuning the coefficients (i.e., weights) attached to each node, which determine the extent to which it activates the nodes in the following layer. Deep neural networks are useful precisely when we have a lot of independent variables associated with a dependent variable through a complex non-linear function. If you can build a simple model with a few variables, then precedence should be given to simplicity. However, deep neural networks are the winners when dealing with complex systems, such as human language. In sum, deep neural networks are models with a specific architecture and a learning process optimized for predicting a dependent variable. They have been used for visual representations, but for textual data, success has been achieved with LLMs.

3.2 LLMs are Probabilistic Guessing Machines

LLMs, like the one you use with ChatGPT, are deep neural networks trained on massive textual datasets to guess masked words. They are deep neural networks with a unique transformer architecture.[2,3] Let me explain this remarkable idea using a simple example.

Imagine a romantic scene in an old movie. A couple is sitting on a beautiful beach. It is an evening with a great atmosphere. They are holding their glasses of wine, looking at each other with blurry eyes. The man says softly: "Will you marry me?" and continues, "I love …". It is an old movie. The soundtrack is polluted with noise, and you cannot hear the last word in the utterance. Can you guess the missing word in the utterance "I love …"? The chances are that it is "you." Is it the only possible word that can complete the sentence? Probably not. However, it is the most likely word. Let us further assume that the soundtrack is back, and when you hear the man's sentence, it is: "I love spinach." In this case, you may be surprised… It may be that the character you just observed is the child from Bateson's famous anecdote …

LLMs are deep neural networks trained on numerous sequences to guess missing words. The independent variables are the words in a sequence, and the dependent variable is masked (i.e., a missing word). However, LLMs are not just trained on words; they are trained on *sequences* of words by paying attention[4] to context. For example, let us provide our model with two different contexts for guessing the missing word. Here are the instructions that I gave to ChatGPT:

---

[2]https://www.youtube.com/watch?v=wjZofJX0v4M&list=WL&index=14.

[3]https://www.youtube.com/watch?v=wjZofJX0v4M&list=WL&index=14.

[4]https://www.youtube.com/watch?v=fjJOgb-E41w&list=WL&index=17.

I am giving you a context. Guess the missing word. The context: John is in the restaurant. The waitress arrives, and John orders a steak and says: "I love [MASKED]."

ChatGPT: Based on the context, John likely says, "I love steak."

Now, I gave ChatGPT a different context:

A couple is sitting on a beautiful beach. It is an evening with a great atmosphere. They are holding their glasses of wine, looking at each other with blurry eyes. The man says softly: "Will you marry me?" and continues, "I love [MASKED]."

In this context, ChatGPT makes a different guess:

In this romantic and intimate setting, the man likely says, "I love you."

We can see that ChatGPT is sensitive to context, and to understand the transformer mechanism underlying this remarkable property, let us move on to the next section.

3.3 LLMs are Contextual Machines

To understand LLMs as contextual machines, we must understand two ideas: *embedding* and *attention*. In the mind of ChatGPT, words are not represented as words. Each word in a sequence is represented as a vector (i.e., an array) of numbers. This vector of numbers, or the "embedding," represents the words co-located with our target words in a large corpus of language and the weight assigned to these co-locations. In other words, the meaning of each word is represented by its context, which is the semantic field in which the word is located. Let me give you an oversimplified example. Imagine a world with only four words: hospital, nurse, pizza, hamburger. We analyze all existing texts and measure the probability that each word appears to the right or the left of each of the other words. We get the following vectors for "hospital" and "pizza":

	Hospital	Pizza
Hospital		0.1
Nurse	0.8	0.1
Pizza	0.1	
Hamburger	0.1	0.8

We can see that the word embedding of "hospital" is strongly connected to "nurse." In contrast, the word embedding of "pizza" is strongly connected to "hamburger." The meaning of a word is represented as an array of numbers indicative of the semantic field in which the word resides.

Word embeddings are, therefore, a type of word representation that allows words to be represented as vectors in a continuous vector space. The main idea behind word embeddings is to represent the semantic field of words in such a way that words with similar meanings (i.e., fields) are located near one another in a vector space. This is achieved through algorithms that process text and learn the lexical field in which words appear. The resulting embeddings capture nuanced relationships between words based on their usage patterns in the text. Let me explain this point using the sequence: "Will you marry me? I love [MASKED]."

Consider the word "marry" as an example to illustrate the concept of word embeddings. In the vector space, the embedding for "marry" would be positioned near other words related to marriage, love, and relationships, such as "wedding," "spouse," "engagement," etc. This proximity in the vector space reflects the contextual and semantic similarities between these words. The embedding of the word "marry" is a multi-dimensional vector that encapsulates its semantic meaning by learning from the contexts in which "marry" appears across a large text corpus. When analyzing a large corpus of texts, we may notice that "marry" is surrounded by words such as marriage, wedding, love, etc. The embedding of the word "marry" is, therefore, an array of words, where words such as those mentioned above (e.g., love) are given more weight than others.

At this point, we understand that each word in the sequence "Would you marry me? I love [MASKED]" is represented as a vector of words. When learning to guess the next word or the masked words elsewhere in the utterance, the neural network learns to pay attention to different words in the sequence and to update the representation of their vector by adjusting them to the vectors of the other words that appear in the same context. For example, let us assume that our network is struggling to decide which word is the best guess for the missing word: "you" or "spinach." To make a decision, it seems that the context words "marry" and "love" are much more informative than "would" and "I." Attention is given to the more informative word embeddings, and within them, to words reflecting similar semantic contexts in other words. Think about the word "love." The words associated with "love" can include words from different semantic fields: other people (e.g., I love Mom), food (e.g., I love wine), etc. Therefore, the word embedding of "love" includes a mixture of weights, indicating how "love" is co-located with other words from different contexts. However, when "love" and "marry" appear in the same sentence, their embedding is updated to give more weight to words reflecting the embedding of the other word. Therefore, the embedding of "love" is updated so that words relevant to the love of *others* gain more weight than words associated with the love of things in other semantic fields, such as food. The final guess produced by the machine is a distribution of the probabilities of words indicating which word is the best guess for completing our sentence. The attention given to the words in context is a key to understanding ChatGPT as a contextual machine. ChatGPT does not simply guess the most probable word, but guesses the next word by paying attention to the context.

For ChatGPT, context is reflected in several senses. First, the representation of words is contextual in the sense that the meaning of each word is captured through its semantic field. Second, context is expressed by the mechanism of considering the other words that appear with each word in the given sentence. This is the context of the here and now. The third aspect of the context is the process through which the embedding of each word is updated by considering the context of the other words in the sentence. The LLMs used in this book are deeply contextual.

3.4   From José Ortega y Gasset to ChatGPT

In his seminal paper "The Difficulty of Reading," Ortega y Gasset [8, p. 14] explains that: "Context is a dynamic whole on which each part exercises influence, modifies the others, and, vice versa, receives pressure from the others." This explanation is strikingly similar to some attempts to explain LLMs using a physics perspective.[5] According to this view, the tokens processed by LLMs like ChatGPT can be thought of as particles. As explained in the article by the physicist Tim Lou (see the last footnote):

"Transformer based language models are treating words as particles, which move around under the influence of each other, generating intriguing patterns."

In LLMs, the tokens are the pieces composing the sequence. These tokens are "particles" influencing each other through a mechanism of attention, or more accurately, "self-attention." Context is, therefore, not only the tokens preceding or surrounding the guessed word but the updated representations of the tokens as a result of their self-attention mechanism. Therefore, it is important to understand that LLMs are contextual machines in which the context is an emergent phenomenon reflecting the "dynamic whole" of the representations and connectivity of those tokens. However, LLMs are different from our minds. For example, they are token-based, while our minds are concept-based. This does not mean that, on the micro-level, our minds are not grounded in tokens. It just means that our cognition is basically conceptual [9] and that we must bridge a communication gap to communicate with a token-based machine. To get these machines closer to human understanding, which is deeply conceptual, we must probe them for a higher level of abstraction. In addition, they are highly sophisticated but "myopic" (i.e., short-sighted) guessing machines. To assist us in understanding context, we must carefully convey our specific needs to the machine. This process is far from trivial. It is like functioning as a translator mediating between two different cultures. This art of translation is the art of "prompting."

---

[5]https://medium.com/towards-data-science/understanding-large-language-models-the-physics-of-chat-gpt-and-bert-ea512bcc6a64.

3.5   The Art of Prompting

In the context of using ChatGPT or similar LLMs, a "prompt" or "prompting" refers to the text input given by a user that instructs or guides the model to generate a response or output. Prompts can range from simple questions, like asking for information or a definition, to more complex requests, such as writing stories, explaining concepts, or generating creative content. The effectiveness and relevance of the LLM's response often depend on the clarity, specificity, and detail of the prompt provided by the user.

Prompting is crucial to interacting with language models because it is the primary communication channel between the user and the LLM. It can be considered the starting point or trigger for the AI's text-generation process. Here are a few critical aspects of prompts as proposed by ChatGPT:

1. **Specificity**: A prompt's level of detail can significantly influence the quality of the AI's response. More specific prompts tend to yield more accurate and relevant answers.
2. **Instructional**: Prompts can include instructions or commands guiding the AI to perform a specific task, such as writing in a particular style, answering a question, or following a set of constraints in its response.
3. **Contextual**: Sometimes, prompts provide context or background information that helps the AI understand the scope or domain of the question or task, leading to better informed and more appropriate responses.
4. **Creative**: In creative tasks, prompts serve as a source of inspiration or a creative brief for the AI, outlining the elements or themes to be included in the generated content.

Let me explain the general structure of prompts by using a simple example. I want to identify the social emotion expressed by an utterance. First, I use a title:

[Identifying the emotion in an utterance]

Next, I briefly explain social emotions, thereby providing ChatGPT with the context for understanding my instructions:

Social emotions arise from interactions with others or social situations. They include empathy, shame, guilt, pride, and jealousy. These play a crucial role in regulating social behavior and maintaining social bonds. Such emotions help individuals navigate complex social landscapes by motivating behaviors that align with societal norms and expectations.

The third step is to give ChatGPT the specific task:

Identify which <SOCIAL EMOTION> is expressed by the following utterance: My son won the first prize in the competition.

The placeholder "<>" is used to direct the machine to a variable with several parameters, as shown below.

Next, I provide ChatGPT with several examples.

Finally, I ask for the form of the output:

Output in the form of a single emotion.

The final prompt appears below with hashtags separating different parts of the prompt:

Prompt 3.1
[Identifying the emotion in an utterance]

Social emotions arise from interactions with others or social situations. They include empathy, shame, guilt, pride, and jealousy. These play a crucial role in regulating social behavior and maintaining social bonds. Such emotions help individuals navigate complex social landscapes by motivating behaviors that align with societal norms and expectations.

##

Identify which <SOCIAL EMOTION> is expressed by the following utterance: My son won the first prize in the competition.

##

Example 1.

Utterance: Thank you for your kindness.

Social emotion: Gratitude.

Example 2.

Utterance: I lied to my best friend.

Social emotion: Shame.

##

Output in the form of a single emotion.

##

<SOCIAL EMOTIONS>: <PRIDE, SHAME, GUILT, REGRET, GRATITUDE>

In sum, the concept of prompting is fundamental to using LLMs, as it shapes the interaction and determines the outcome of the text generation process. Although the literature discusses the idea of prompting in terms of "prompt engineering," effective prompting is both an art and a science, and users often develop strategies to craft prompts that lead to the most meaningful, accurate, or creative responses from the LLM. Here, I must emphasize the importance of *domain expertise*. Understanding context is not an aim in itself. It is one crucial aspect of solving a problem in a specific domain. Designing a prompt is impossible without understanding the domain, your task, and the way to ask the right questions and provide

ChatGPT with the best instructions. I present many prompts in the following chapters, specifically those presenting worked examples. These prompts rely on expertise and experience, and none emerged with no efforts. All of them involve a learning process, sometimes more frustrating than expected. When moving to real-world challenges, no shortcuts are usually available.

References

1. Starmer, J.: The StatQuest Illustrated Guide to Machine Learning!!! StatQuest Publications Inc., USA (2022)
2. Goodfellow, I., Bengio, Y., Courville, A.: Deep Learning. MIT Press, Cambridge (2016)
3. Skansi, S.: Introduction to Deep Learning: From Logical Calculus to Artificial Intelligence. Springer, New York (2018)
4. Prince, S.J.: Understanding Deep Learning. MIT Press, Cambridge (2023)
5. Amaratunga, T.: Understanding Large Language Models: Learning Their Underlying Concepts and Technologies. Apress, Berkeley (2023)
6. Zhao, W.X., Zhou, K., Li, J., Tang, T., Wang, X., Hou, Y., Wen, J.R.: A survey of large language models. arXiv preprint arXiv:2303.18223 (2023)
7. Wolfram, S.: What Is ChatGPT Doing … and Why Does It Work? Wolfram Inc., New York (2023)
8. Gasset, O.: The difficulty of reading. Diogenes 7(28), 1–17 (1959)
9. Harnad, S.: To cognize is to categorize: cognition is categorization. In: Handbook of Categorization in Cognitive Science, pp. 21–54. Elsevier, New York (2017)

Chapter 4
The One-Hundred Psychological Dimensions of Context

Abstract As a first example of contextual analysis, I present an approach we developed to measure one hundred situational dimensions. These situational dimensions describe how the individual represents the situation in which they are involved. This approach is illustrated using a heated conversation between a school principal, a student, and the student's mother.

4.1 The Psychology of Situations

One of my papers presented an approach for AI-driven context analysis [1], where I adopted a psychologically oriented approach to situations developed by Funder. From a psychological perspective, situations involve the way individuals *represent* the context in which they operate. This perspective focuses on the *individual*, but it does not mean that it falls into the pitfall of idiosyncratic thought, according to which each of us holds a personal and non-communicative representation of situations.

The evolutionary perspective suggests that, beyond cultural and individual differences, the representations of situations may have a common denominator grounded in our basic human experiences. For example, regardless of cultural or individual particularities, representing a situation as "potentially enjoyable" is a level of representation shared by different human beings regardless of their unique perspective. One person may identify a family gathering as potentially enjoyable, while the other may similarly identify an evening in their favorite BDSM club. The fact that different people may identify different situations as potentially enjoyable does not invalidate the existence of "potentially enjoyable situations." Different people may hold different views about what an enjoyable situation is. However, they share a basic sense of joy when describing a situation as enjoyable.

One possible way of understanding context from a psychological perspective is to use the Riverside Situational Q-Sort (RSQ), a questionnaire measuring situational representations by the individual [2]. The RSQ was developed in correspondence with the California Adult Q-Sort (CAQ), which is a set of statements

© The Author(s), under exclusive license to Springer Nature Switzerland AG 2024
Y. Neuman, *AI for Understanding Context*, SpringerBriefs in Computer Science,
https://doi.org/10.1007/978-3-031-64210-4_4

used for personality assessment [3]. The items in the CAQ are personality descriptions, such as:

X is a talkative person.
X is basically anxious.
X has hostility toward others.

When developing the RSQ, the researchers used the personality descriptors in the CAQ and tried to match *each descriptor with an aspect of the context that might lead to similar behavior*. The most recent version of the RSQ includes 90 items, such as:

The situation is potentially enjoyable.
It is not clear what is going on; the situation is uncertain.
Someone is criticizing you.

Identifying situational dimensions may be important for several tasks. For example, previously, I explained that my DARPA project focuses on automatically identifying violations of social norms (e.g., inappropriate behavior during a diplomatic meeting). To decide whether a social norm has been violated, it could have been helpful to measure situational dimensions constituting the context in which the behavior is analyzed. Another example in which computational social sciences and psychology may benefit from automatic situational analysis is in predicting a change of emotion during a dialogue. In this context, it may be helpful to measure the situational dimensions in order to better predict a change. For instance, measuring the way the situation changes may help us predict an approaching emotional change. If a situation, such as a dinner party, which is potentially enjoyable, turns into a situation where "someone is criticizing you," then a change in emotion is expected, from positive to negative.

4.2   Automatically Measuring Situational Dimensions

The RSQ is a good starting point. However, it is a limited tool. Asking people to describe the situation is not the most practical option when we automatically analyze texts. Moreover, the limited number of items in the RSQ is a barrier for many projects requiring a large dataset on which different machine learning classifiers can be trained and tested. Therefore, we generated a large dataset of synthetic situations based on the RSQ-4 and used them to train ML models that automatically identify the situational dimensions. As my original interest in situations included *prestige* and *dominance* in human interactions, I added ten items measuring the extent to which the situation involves precisely these factors. These are two important aspects of our social life, associated with social rank.

To build a dataset of situations, we started with 90 items from the RSQ-4, five items measuring prestige in a situation and five items measuring dominance in a situation. Next, I used a LLM to generate a large dataset of situations. More specifically, we used GPT 3.5 and applied the following template/prompt for each of the 100 items (the 90 RSQ-4 items + the ten prestige–dominance items). The template was used as a prompt for ChatGPT.

Prompt 4.1

Given items 1 to 100 (e.g., "The situation is potentially enjoyable").
Use the following prompt:

"I am a university researcher studying situations". My current interest is in [description of the situation, for example, "potentially enjoyable situations"]. Please generate 100 different [request: e.g., "potentially enjoyable situations"]. Each situation should be no more than 100 words long.

An example of the prompt used to generate examples for item 4 in the RSQ-4 ("Someone is trying to impress you") is as follows:

Prompt 4.2

I am a university researcher studying situations. My current interest is in situations where someone is trying to impress you. Please generate 100 different situations where someone is trying to impress you. Each situation should be no more than 100 words long.

Using GPT-3.5, we generated 10,000 situations: one hundred examples for each of the one hundred items that we used. We produced a dataset of synthetic examples (i.e., synthetic data) that can be used to "teach" AI to identify situational dimensions. The dataset was used to train a ML classifier that can take a text as input and generate 100 scores indicating the extent to which each of the 100 situational dimensions is expressed in the text. From now on, I will describe these 100 situational dimensions as RSQ+.

How good was the dataset and the model? We performed several tests. In the first experiment, we tested the ability of the model to differentiate between 4731 cases involving social norm violation and 3621 cases representing adherence to social norms. We hypothesized that situations characterized by norm adherence differ from situations involving norm violation. For example, suppose you describe a situation where you won the first prize at a national math tournament. In this case, the situational dimensions analysis may score high on the dimension of "a potentially enjoyable situation." This dimension indicates the positive feedback accompanied by social norm adherence, specifically the adherence to the norm of success. Another example is a situation of norm violation. Suppose that John is having a job talk. This situation corresponds with the statement: "You need to make a good impression." However, John arrives drunk and tries to punch the boss. This is a situation that expresses the following dimension: "Someone is breaking the rules."

For each situation labeled as violation or adherence, we generated the 100 RSQ+ scores and used these one-hundred scores as input features for training another ML classifier. This classifier successfully identified norm violations and norm adherence. In our dataset, 52% of the cases were labeled as violations. The classifier correctly identified 86% of them with 94% precision. The other 48% of the cases were instances of adhering to a social norm. Using the RSQ+ dimensions, the classifier correctly identified 95% of them with 87% precision. We can see that the psychological representation of situations can be successfully used to understand whether the situation involves behavior that expresses an adherence to a social norm or a violation of a social norm.

4.3 The Psychological Context of a Conversation

In our original paper, we trained a model on a synthetic dataset of examples. However, we can use ChatGPT directly to identify the RSQ+ situational dimensions. For this task, I created a revised version of the 100 statements that better fit the prompt given to ChatGPT. From now on, I will describe the tool as *100PCD* (i.e., 100 psychological contextual dimensions). Here is an illustration of how I used the tool. The context is an imaginary conversation between Pavel (P), a school principal, a school pupil by the name of Volodya (V), and his mother, Maria (M). The principal has invited the mother and son to discuss Volodya's behavior. The conversation is as follows:

Maria: Hello, Principal Pavel. May we come in?
School Principal Pavel: Hello Maria. Please have a seat.
Maria: Thank you.
School Principal Pavel: Volodya misbehaved in school. He vandalized school property.
Maria: Please accept my apology for my son's behavior.
School Principal Pavel: Volodya, you behaved like a criminal and not for the first time.
Volodya: Fuck you, Principal Pavel, I don't care what you think.
Maria: Volodya, please calm down.
Volodya: I will not be calm!
School Principal Pavel: And sit back on your chair!
Volodya: I will not!
Maria: Please, calm down, please.
School Principal Pavel: Maria, you are a very nice and intelligent woman, but your son is a criminal.
Maria: Oh, Principal Pavel … Can we discuss the possibility of just giving a suspension?
School Principal Pavel: I am considering kicking him out of school. Suspension is not an option.
Volodya: Why don't you send me to jail?!
School Principal Pavel: Your proposal is accepted. Now get out of my office!
Volodya: You will regret this and you will see that one day I will be an important person, maybe the president of Russia.

Here is the prompt that I used for analyzing the conversation:

Prompt 4.3

The Riverside Situational Q-Sort (RSQ) is a method used in psychology to assess an individual's perception of their social environment. It involves sorting a set of statements or items based on how accurately they describe the individual's current situation or circumstances. The RSQ helps psychologists understand how people perceive and interpret their social context, which can provide insights into their thoughts, feelings, and behavior in different situations.

##

Here are the 100 statements composing the revised RSQ [100 STATEMENTS HERE]:

##

Describe which of the above-mentioned 100 RSQ statements best describes the conversation [CONVERSATION HERE].

ChatGPT produced the following output:

Based on the provided list of statements from the Riverside Situational Q-Sort (RSQ), as I revised them to the specific task, the statement that best describes the conversation is:

It is a situation where people who are present occupy different social roles or levels of status.

As ChatGPT explains, this statement reflects the diverse range of topics and roles in the conversation, indicating that the participating individuals may have varied social roles, outlooks, or levels of influence. This output grasps an important aspect of the conversation, namely that it involves a context with people of *different social statuses*. Do you remember Bateson and the idea of context and alternatives? What should be expected from a conversation involving people with different social statuses? One clear expectation concerns *power relationships*. People with a higher social rank are expected to express dominance, and those with a lower social rank are expected to show compliance. Therefore, we expect Principal Pavel, who holds a higher social rank, to express dominance and Volodya, the school student with the lower social rank, to express compliance. Is this the case? The answer is negative, and therefore, we may easily conclude that Volodya's behavior violates our expectations and can be identified as violating a social norm.

Volodya's behavior can also be described as an example of *markedness*. As explained before, the least expected behavior is the most informative. Nothing surprises us about Principal Pavel's behavior. The principal is an authority with a higher social status and expresses his status and dominance. Volodya surprises us by violating our expectations and showing behavior that is of interest.

One way to design a system that can pick out social norm violations is to identify the situation and test whether the participants' behavior violates or adheres to some expectations resulting from the situation. I will elaborate on this idea in the case studies.

We can further analyze the conversation with a higher level of resolution by focusing on each utterance and instructing ChatGPT as follows:

Prompt 4.4

For each utterance in the following conversation, describe which of the above-mentioned 100 RSQ statements best describes it.

The output produced by ChatGPT was:
Maria: Hello, Principal Pavel. May we come in?

- RSQ Statement: "It is a situation where someone asks for something."

School Principal Pavel: Hello Maria. Please have a seat.

- RSQ Statement: "It is a situation where talking is permitted."

Maria: Thank you.

- RSQ Statement: Not applicable.

School Principal Pavel: Volodya misbehaved in school. He vandalized school property.

- RSQ Statement: "The situation involves someone blaming another for something."

Maria: Please accept my apology for my son's behavior.

- RSQ Statement: "It is a situation where someone needs help."

School Principal Pavel: Volodya, you behaved like a criminal and not for the first time.

- RSQ Statement: "It is a situation where someone is criticizing."

Volodya: Fuck you, Principal Pavel, I don't care what you think.

- RSQ Statement: "It is a situation where someone attempts to dominate or boss others."

Maria: Volodya, please calm down.

- RSQ Statement: "It is a situation where someone needs or desires reassurance."

Volodya: I will not be calm!

- RSQ Statement: "It is a situation that could make people tense and upset."

School Principal Pavel: And sit back on your chair!

- RSQ Statement: "It is a situation where someone is attempting to dominate or boss others."

Volodya: I will not!

- RSQ Statement: "It is a situation where someone is attempting to dominate or boss others."

Maria: Please, calm down, please.

- RSQ Statement: "It is a situation where someone needs or desires reassurance."

School Principal Pavel: Maria, you are a very nice and intelligent woman, but your son is a criminal.

- RSQ Statement: "It is a situation where someone blames another for something."

Maria: Oh, Principal Pavel … Can we discuss the possibility of just giving him a suspension?

- RSQ Statement: "It is a situation where someone needs help."

School Principal Pavel: I am considering kicking him out of school. Suspension is not an option.

- RSQ Statement: "It is a situation where someone is attempting to dominate or boss others."

Volodya: Why don't you send me to jail?!

- RSQ Statement: "It is a situation that could be intellectually stimulating."

School Principal Pavel: Your proposal is accepted. Now get out of my office!

- RSQ Statement: "It is a situation where someone attempts to dominate or boss others."

Volodya: You will regret this and you will see that one day I will be an important person, maybe the president of Russia.

- RSQ Statement: "It is a situation where ambition can be expressed or demonstrated."

ChatGPT is not always correct, specifically when it concerns the sophistication of language, such as the ironic comment by Volodya. However, beyond the particularity of each statement, we can identify *general patterns of interaction* and behavior using the following prompt:

Prompt 4.5
Please identify the RSQ statements that best capture the predominant emotions, behaviors, and roles exhibited by each participant in the conversation, highlighting the common themes that characterize their interactions.

ChatGPT responded as follows:
Let us analyze the conversation and identify the RSQ statements that best capture the predominant emotions, behaviors, and roles exhibited by each participant:

Maria:

- RSQ Statements:
- "It is a situation where someone is asking for something." (Initiating the conversation)
- "It is a situation where someone needs help." (Apologizing for her son's behavior and seeking assistance)
- "It is a situation where someone needs or desires reassurance." (Attempting to calm Volodya down and seeking reassurance)
- "It is a situation where someone is blaming another for something." (Apologizing for her son's behavior)
- **Common Themes**: Seeking assistance, expressing concern for her son's behavior, attempting to calm the situation down.

Volodya:

- RSQ Statements:
- "It is a situation where someone attempts to dominate or boss others." (Challenging the authority of Principal Pavel)
- "It is a situation that could make people tense and upset." (Expressing frustration and agitation)
- "It is a situation where ambition can be expressed or demonstrated." (Asserting his future ambitions)
- **Common Themes**: Defiance against authority, emotional agitation, expression of ambition.

School Principal Pavel:

- RSQ Statements:
- "It is a situation where someone is criticizing." (Critiquing Volodya's behavior)
- "It is a situation where someone attempts to dominate or boss others." (Exerting authority over Volodya)
- "It is a situation where someone needs or desires reassurance." (Addressing Maria's concern with reassurance)
- **Common Themes**: Exercising authority, addressing misconduct, and attempting to control the situation.

Common Themes across Interactions:

- RSQ Statements:
- "It is a situation where someone is attempting to dominate or boss others." (Evident in interactions between Volodya and Principal Pavel)
- "It is a situation where someone needs or desires reassurance." (Evident in Maria's attempts to calm Volodya down and seek reassurance)
- "It is a situation where someone is blaming another for something." (Evident in Maria apologizing for her son's behavior)

- **Common Themes**: Power dynamics, struggles for authority, emotional tension, and seeking reassurance and resolution.

The common themes identified by ChatGPT provide a valid representation of the conversation. For instance, we can see that ChatGPT correctly identified the *power dynamics* we previously identified at the macro level of the conversation. Representing the conversation using the situational dimensions of the *100PCD*, we can identify psychological situational dimensions that may be used to understand the participants better and anticipate their behavior. For example, it is clear that by going against authority, Volodya presents a rebellious personality. Holding a higher social status, Principal Pavel is expected to respond angrily to such a challenge. His anger expresses the threat to his status. This contextual situational analysis is automatic and can be used for various AI tasks.

Consider, for example, a married couple participating in a family therapy session. Long ago, John Gottman [4] identified couple dynamics that can predict divorce. His approach was tested long before the AI era. Now, think about AI "listening" to the conversation and analyzing it using the situational dimensions of the *100PCD*. The system can quickly identify when the couple is entering the dangerous zone of a power struggle, bringing to their attention the fact that they are engaging in a destructive process. In the first phase of the meeting, where the husband and wife present the background and reasons for seeking therapy, the AI may also gain important contextual information that can help it to anticipate the couple's moves. In sum, using a simple tool that can identify the psychological dimensions of a situation, we can gain important contextual knowledge that can help us to understand the conversation and the participants. As illustrated above, these dimensions can be further used for prediction and control.

In sum, this chapter presents one basic approach for representing, measuring, and using contextual situational dimensions, theoretically motivated by a specific psychological approach. In the next chapter, I present a more general and comprehensive approach, clearly influenced by the ideas presented previously by Voloshinov, Bateson, and Becker.

References

1. Neuman, Y., Cohen, Y.: A dataset of 10,000 situations for research in computational social sciences, psychology and the humanities. Sci. Data **10**(1), 505 (2023)
2. Funder, D.C.: Taking situations seriously: the situation construal model and the Riverside Situational Q-Sort. Curr. Dir. Psychol. Sci. **25**, 203–208 (2016)
3. Lanning, K., Sherman, R.: The California Adult Q-Sort. In: Zeigler-Hill, V., Shackelford, T.K. (eds.) Encyclopedia of Personality and Individual Differences. Springer, Berlin (2017). https://doi.org/10.1007/978-3-319-28099-8_11-1
4. Gottman, J.M.: What Predicts Divorce? The Relationship Between Marital Processes and Marital Outcomes. Psychology Press, UK (2014)

Chapter 5
Conceptualizing Context
for Computational Tasks

Abstract I present an integrative approach for analyzing context. First, I suggest differentiating between a situation and a context. I present seven dimensions of a situation (e.g., the ambiance) and four dimensions of a context (e.g., the shared knowledge needed to understand the situation). This generic approach is explained and illustrated. The approach will be used in the following chapters to guide the analysis of various situations.

5.1 Defining Context

In this chapter, I present an operational definition of context. This means that I explain how to "measure" context. The definition aims to support the analysis of textual data, where "textual" means any set of written or spoken signals used for social communication. Integrating previous ideas, I propose to differentiate between two concepts: *situation* and *context*.

A situation involves the here and now. It describes the people involved (who) and their intent (why), the place (where) and time (when), objects present in the situation, the ambiance or atmosphere present in the situation (ambiance), and procedures or actions that were taken (how). Being here and now does not mean a situation lacks a general structure. The situation of a *dinner party* materializes in particular dimensions. However, it is also characterized by a general structure guiding our *general* expectations from a dinner party.

The context describes the text's *coherence* in terms of inferences and hypotheses that can be drawn from it and the relation between its parts, the *shared knowledge* required to understand the situation, the *expectation* of those involved in the situation, such as anticipated responses to utterances, and the *common evaluation* of the situation by the participants in terms of basic sentiments.

5.2   The Dimensions of a Situation

I start with the idea of a *situation*. The general definition of a situation is the *specific circumstance* or events within a context. This definition is uninformative. Therefore, I identify the following seven dimensions of a situation:

1. WHO: The individuals involved in the interaction.
2. WHY: The underlying motivation or intention of each individual.
3. WHERE: The physical location of the interaction.
4. WHEN: The specific time or period when the interaction occurs.
5. OBJECTS OR WHAT: The tangible or intangible objects present or relevant within the interaction.
6. HOW: Information about the procedure characterizing the situation.
7. AMBIANCE: The atmosphere that characterizes the situation.

I will use slightly different ways of conceptualizing a situation in different contexts of analysis (e.g., conversation vs. a short story), but the general scheme will be the same. Let me illustrate these dimensions and how they can be extracted from the conversation between Pavel, Maria, and Volodya. First, I identify the *ambiance* using the following prompt:

Prompt 5.1
Ambiance is the mood, character, quality, tone, atmosphere, etc., particularly of an environment or milieu.

##

Describe the ambiance expressed in the following conversation: [CONVERSATION HERE]

##

Output is in the form of bullet points with one word for each bullet point (e.g., SERENE, WARM).

The output produced by ChatGPT is as follows:

- TENSE
- CONFRONTATIONAL
- DISAPPOINTED
- DEFIANT
- STRESSFUL

We can see that ChatGPT validly represents the atmosphere of the conversation.

To extract the other dimensions, I use the following prompt:

Prompt 5.2

Extract from the following conversation the situational dimensions <WHO> <WHY> <WHEN> <WHERE> <OBJECT> <HOW>.

##

Provide your best guess if a <SITUATIONAL DIMENSION> is not explicitly mentioned.

##

The conversation is [CONVERSATION HERE].

##

<WHO>: The individuals involved in the conversation.
<WHY>: The motivation or intention underlying the behavior of each individual.
<WHERE>: The physical location of the conversation.
<WHEN>: The specific time or period where the conversation occurs.
<OBJECTS or WHAT>: The tangible or intangible objects present or relevant within the conversation.
<HOW>: Information about a certain procedure.
<SITUATIONAL DIMENSION>: <WHO>, <WHY>, <WHEN>, <WHERE>, <OBJECTS>, <HOW>.

##

OUTPUT: A list of parameters for each <SITUATIONAL DIMENSION>.

ChatGPT produced the following output:

<WHO>:
Maria
School Principal Pavel
Volodya
<WHY>:
Maria apologized for her son's behavior and tried to calm the situation.
School Principal Pavel: Addressing Volodya's misbehavior, considering disciplinary action.
Volodya: Displaying defiant and aggressive behavior, resisting authority.
<WHEN>:
Not explicitly mentioned.
<WHERE>:
School Principal Pavel's office.
<OBJECTS or WHAT>:
School property (mentioned in the context of vandalism).
Chairs (mentioned in the context of sitting down).

<HOW>:
Maria and Volodya engage in verbal communication with the school principal.
<SITUATIONAL DIMENSION>:
<WHO>: Maria, School Principal Pavel, Volodya
<WHY>: Maria apologizing, school Principal Pavel addressing misbehavior, Volodya displaying defiance
<WHEN>: Not specified
<WHERE>: School Principal Pavel's office
<OBJECTS or WHAT>: School property, chairs
<HOW>: Verbal communication

My interest is in extracting the general dimensions, and therefore, I would like to abstract the general social roles of the individuals participating in the exchange:

Prompt 5.3
Please guess the social roles of <WHO>.
Output: A minimal description of the social role, using no more than three words.

ChatGPT produced the following output:

<WHO>:
Maria: Parent
School Principal Pavel: School administrator
Volodya: Student

So, on a higher level of abstraction, we observe an interaction between a school administrator, a parent, and a student in the principal's office. I would also like to identify the more abstract category of the objects that we observe in the situation, and therefore, use the prompt:

Prompt 5.4
For <OBJECTS or WHAT>, provide the semantic level or category that best represents it according to WordNet and Eleanor Rosch's categorization theory.

This prompt aims to transform some of the output into a higher level of abstraction. According to Eleanor Rosch's categorization theory, the optimal level of categorization is the *basic level*. This level of categorization is neither too abstract nor too concrete, but rather, it represents a level of specificity that is psychologically salient and commonly used in everyday communication. For example, when categorizing objects, we tend to use basic-level categories such as "bird," "chair," or "car." These categories are more specific than extensive categories like "animal," "furniture," or "vehicle," but they are also less specific than more detailed categories like "raven," "armchair," or "sedan." Rosch argued that basic-level categories are *optimal* because they provide enough information to be informative and recognizable while also being broad enough to encompass a variety of instances within the category. Basic-level categories are also typically the first categories children learn and are the most commonly used in everyday conversation, suggesting their importance in cognitive processing and communication.

Rosch's idea of basic-level categories is highly relevant for many computational tasks in psychology and the social sciences, where we are looking to represent the dimensions using an optimal level of abstraction.

ChatGPT produced the following output:

<OBJECTS or WHAT>:
School property: Tangible objects

Identifying the chairs as "tangible objects" is not informative. We may learn that abstracting some objects in the situation requires a better prompt considering the objects' given level of abstraction. If an object is already represented optimally then asking ChatGPT to represent is optimally might confuse the machine.

So far, we have automatically extracted the situational dimensions of the conversation. Next, we should ask whether identifying the parameters of the situation can help us understand the interaction better. The *100PCD* situation analysis taught us that the conversation between Pavel, Maria, and Volodya is characterized by power dynamics, struggles for authority, emotional tension, and attempts to obtain reassurance and resolution. Let me explain the idea of power struggles and how they might violate social norms.

Power relationships express one party's dominance, control, influence, and authority over the other. In some contexts, the power relations are clear, and if they are challenged, it is implied that a social norm has been violated. To identify power relationships, we should determine who is dominant and who is submissive in the conversation. The default is to hypothesize that Principal Pavel is the authority regarding power dynamics, and Maria and Volodya should respond to his utterances by expressing compliance. Let us recall the beginning of the conversation:

Maria: Hello, Principal Pavel. May we come in?
School Principal Pavel: Hello Maria. Please have a seat.
Maria: Thank you.
School Principal Pavel: Volodya misbehaved in school. He vandalized school property.
Maria: Please accept my apology for my son's behavior.
School Principal Pavel: Volodya, you behaved like a criminal and not for the first time.
Volodya: Fuck you, Principal Pavel, I don't care what you think.

You can see that the hypothesis is confirmed for Maria. Whenever Principal Pavel addresses her, she responds with compliance. However, Volodya's response is confrontational and violates our expectations/hypothesis. As the social norm suggests that a response to authority should not be confrontational, we could have expected Volodya to respond with an apology, for instance. However, Volodya's response is confrontational and expresses markedness. Something interesting and informative has just happened. Volodya's response is not coherent with the social rules of the participants in the conversation, and it indicates something interesting about Volodya. To understand this point further, we now turn to context.

5.3 The Dimensions of Context

Context is the second term that I would like to define. I propose four dimensions of context:

1. **Shared knowledge**: The information, understanding, or experiences commonly held among or known to participants in a conversation.
2. **Coherence**: The relations between the text components expressed in inferences and hypotheses we may generate from the text.
3. **Expectations**: The participants' *expected* response to their behavior.
4. **Evaluation**: How the participants evaluate the situation on a negative–positive scale.

Let me first identify the shared knowledge required to understand the situation:

Prompt 5.5
Shared knowledge describes the *information*, *understanding*, or *experiences* expected from the participants in a conversation. Shared knowledge is necessary to understand the conversation.

##

Describe the shared knowledge required to understand the following conversation: [CONVERSATION HERE].

##

Provide your best guess if the shared knowledge cannot be easily identified.

##

ChatGPT produced the following output:

Information:

School hierarchies and roles: Understanding the roles of a school principal and a student's parent in addressing behavioral issues.

Behavioral standards in schools: Knowledge of what constitutes acceptable vs. unacceptable behavior within a school setting.

Disciplinary actions: Awareness of the range of disciplinary actions schools can take in response to student misconduct, from suspension to expulsion.

Parental responsibility: Recognition of the expectation for parents to apologize or take responsibility for their children's actions in school.

Understanding:

Communication norms: Insight into the formal and respectful way of addressing figures of authority, such as school principals, and the expectation of civility in discussions about serious issues.

Consequences of actions: Understanding that actions, especially negative behavior like vandalism, have consequences that can affect one's future opportunities.

Role of apologies: Understanding the social function of apologies in attempting to mitigate conflicts or disciplinary actions.

Experiences:

Parent–teacher meetings: Familiarity with the context of a parent–teacher (or principal) meeting, typically convened to discuss a student's behavior or academic performance.

School disciplinary processes: Experience or knowledge of how schools handle disciplinary issues, including meetings with students and parents and the potential for severe consequences like expulsion.

Conflict resolution: Experience or knowledge of de-escalating a tense situation, particularly involving figures of authority and young individuals.

Aspirations and defiance in youth: Insight into the psychology of youthful defiance and the expression of grand aspirations as a form of resistance or self-defense against figures of authority.

When we analyze the conversation in which a school principal is meeting a mother and her son, we should use the relevant context that includes our understanding of school hierarchies, parental responsibilities, the consequences of actions, and the role of apology, among other things. Volodya's behavior in school and at the meeting indicates that he violates the norms of hierarchy and school behavior, as well as the norms of communication, apology, and understanding the consequences of misbehavior. Regarding Becker's "pathology" of language, Volodya's behavior indicates a pathology. The pathology can be attributed to:

(1) a lack of understanding of the situational constraints (i.e., the context),
(2) misunderstanding of the situational constraints, or
(3) deliberate violation of the situational constraints.

To identify the evaluation dimension, I use the following prompt:

Prompt 5.6
Please evaluate the sentiment experienced by each character in the above-mentioned conversation and provide one of the following words as the output: POSITIVE, NEGATIVE, or NEUTRAL.

ChatGPT proposed the following output:

Here is an evaluation of the sentiment experienced by each character:

Maria: NEUTRAL
Maria starts with a polite greeting but later attempts to calm the situation, showing concern for her son's behavior. However, her sentiment remains relatively neutral throughout the conversation.

School Principal Pavel: NEGATIVE
Principal Pavel expresses frustration and disappointment with Volodya's behavior. He uses stern language and decides on severe disciplinary action, indicating a negative sentiment.

Volodya: NEGATIVE
Volodya displays defiance, anger, and disrespect towards Principal Pavel. His use of profanity and aggressive behavior indicates a negative sentiment.

Overall, the sentiment experienced by each character is predominantly NEGATIVE.

As we can see, ChatGPT perfectly identified the way each participant evaluated the conversation.

To measure coherence, I used the following prompt:

Prompt 5.7
In understanding a conversation, we can differentiate between inferences and hypotheses.

1. Inferences involve drawing logical conclusions from or interpreting information explicitly stated or implied within the conversation.
2. In a conversation, a hypothesis can be seen as a tentative explanation or interpretation of specific elements or events within the conversation. It involves proposing a possible relationship, explanation, or outcome based on the evidence or information presented in the conversation.

   ##

   Produce inferences and hypotheses for the above-mentioned conversation.

   ##

   Inferences and hypotheses should focus only on the psychological and social aspects of the situation.

   ##

   Output in the form of points briefly describing each inference or hypothesis.
   Describe the response each participant in the above-mentioned conversation is anticipating from each of the other participants he or she addresses.

   ChatGPT produced the following output:

Inferences

- Volodya's habitual misbehavior: The mention of Volodya behaving "like a criminal and not for the first time" implies that this incident is not isolated. It suggests a *pattern of problematic behavior* from Volodya.
- Principal Pavel's authority: The principal's ability to propose expulsion indicates a position of significant authority within the school, responsible for maintaining discipline.

- Maria's concern for Volodya: Maria's attempts to apologize for Volodya and negotiate a less severe punishment suggest *she is concerned about the implications of his actions on his future.*
- Volodya's defiance: His confrontation and refusal to comply with instructions indicate *a deep-seated defiance against figures of authority.*
- Principal Pavel's perception of Volodya: Labeling Volodya as a criminal suggests that Principal Pavel sees him as going beyond simple schoolyard misbehavior, viewing him as a significant discipline problem.

The inferences enrich our understanding of Volodya. A student with (1) a pattern of problematic behavior and (2) *a deep-seated defiance against figures of authority.*

What are the hypotheses generated by ChatGPT? Here they are:

Hypotheses:

- Volodya's behavioral issues stem from external factors: Given the repeated nature of Volodya's misconduct and extreme reaction, his behavior may be influenced by external factors or underlying issues not discussed in the conversation, such as family problems, societal influences, or personal challenges.
- Maria's potential conflict: Maria's polite demeanor and attempts to mitigate the situation suggest she might be in a difficult position, balancing her understanding of her son's wrongdoing with her instinct to protect him. This could indicate a complex family dynamic or her struggle with effective parenting strategies.
- Volodya's provocative statement about becoming president: This could reflect Volodya's deep-seated aspirations or a facet of his defiant personality, suggesting he desires recognition or to prove his worth beyond the current disciplinary context. It may also indicate a mismatch between his self-perception and the way figures of authority perceive him.
- Principal Pavel's decision-making process: His immediate jump from suspension to considering expulsion might indicate previous encounters with similar situations, suggesting possible exhaustion of patience, or resources for dealing with such cases, or a strict policy on vandalism.
- Effect of Volodya's actions on the school environment: Volodya's behavior and the principal's reaction to it could affect the school's atmosphere, possibly creating a more tense or punitive environment. This might influence the way students perceive authority and misconduct.

One interesting hypothesis concerns Volodya's aspiration to become a president. Here, we have a school student with a repeated pattern of misbehavior and confrontational behavior that expresses misunderstanding or refuses to accept contextual constraints and norms of behavior. ChatGPT hypothesizes that Volodya's provocative statement about becoming president may reflect his deep-seated aspirations or a facet of his defiant personality, suggesting that he desires recognition or to prove his worth beyond the current disciplinary context. It may also indicate a mismatch between his self-perception and the way figures of authority perceive him. These hypotheses may further guide our understanding: a young man with a repeated pattern of misconduct, confrontational behavior, and disrespect for social norms aspires to be a president. Is it an aspiration expressing a desire for recognition or a mismatch indicating some personality disorder? We will answer this question when we come to analyze all the contextual dimensions of the conversation.

To identify the expectations, I used the following prompt:

Prompt 5.8

In a conversation, an expectation describes the response each participant expects from an interlocutor to which they address their utterance.

##

Describe the characters' expectations in the following conversation: [CONVERSATION HERE].

ChatGPT proposed the following expectations:

Maria's Expectations

1. **Expectation for entry and audience**: Maria expects Principal Pavel to grant her and her son permission to enter his office and discuss Volodya's behavior.
2. **Expectation for understanding and leniency**: By apologizing for Volodya's actions, Maria likely expects some level of understanding or empathy from Principal Pavel, hoping for a resolution that does not severely impact Volodya's future at the school.
3. **Expectation for dialogue**: When suggesting the possibility of just having a suspension, Maria anticipates a negotiation over the consequences of Volodya's actions, aiming for a less severe punishment.

School Principal Pavel's Expectations

1. **Expectation for acknowledgment and responsibility**: Pavel expects Maria (and possibly Volodya) to acknowledge the seriousness of Volodya's actions and to take responsibility for them, as indicated by his direct statements about Volodya's behavior.
2. **Expectation for behavioral change**: By labeling Volodya's actions and discussing severe consequences, Pavel likely expects a realization of the gravity of the situation from Volodya, hoping it might catalyze a change in behavior.
3. **Expectation for respect and compliance**: As the principal, Pavel expects a certain level of respect and compliance from both Maria and Volodya, which is evident in his directives for Volodya to sit back and his reaction to Volodya's defiance.

Volodya's Expectations

1. **Expectation for autonomy**: Volodya's defiant responses indicate he expects to assert his autonomy, resisting attempts by figures of authority to control his behavior or dictate how he should respond.
2. **Expectation to be heard**: His provocative statement about becoming an important figure suggests Volodya expects his words to carry weight and to be taken seriously, even if they are expressed in a confrontational manner.
3. **Expectation for rebellion**: Given his behavior and responses, Volodya seems to expect that his rebellion against the principal's authority will not have immediate personal consequences that he cannot handle or is willing to face, as seen in his challenge to the principal about being sent to jail.

Let me focus again on Volodya. According to the above analysis, Volodya expects Principal Pavel to accept his behavior even if it is confrontational, and he seems to expect there to be no consequences to his actions. This representation of consequences is a recurring theme in our contextual analysis, and it leads us back to the question I asked at the end of the previous section:

Does Volodya's aspiration to become the president express a desire for recognition or a mismatch indicating some personality disorder?

The contextual analysis provided so far leads us to consider the second hypothesis more probable than the first, a point discussed in the concluding section of this chapter.

5.4 Summary and Conclusions

This chapter presents a theory of context that leans heavily on the ideas previously mentioned. Context is the pattern supporting the entropy reduction of our understanding and expectations. It is composed of situational dimensions and contextual dimensions. We applied this to a specific conversation in which a mother and her son are at the principal's office and the principal addresses the son's misbehavior. The mother apologizes for her son's behavior and tries to calm the situation down. However, the son displays aggressive and confrontational behavior, challenging the school principal's authority. The atmosphere is tense and confrontational. The son evaluates the situation negatively and challenges the norms of social hierarchy. Moreover, he seems to misunderstand or dismiss the consequences of his behavior and the disciplinary power of the principal. The contextual dimensions automatically extracted from the conversation may be used to get a better understanding of each participant's next moves. To the extent that the conversation is our benchmark for understanding the characters, we can guess, following Mischel, that Volodya's rebellious personality will be expressed when interacting with authority figures. In other words, we should expect a personality–context pattern of type

REBELLIOUS–AUTHORITY. Moreover, we may predict that Volodya's "pathology" of misunderstanding contextual constraints indicates a deeper problem than the developmental phase of a teenager trying to express his autonomy.

In the 1955 film "Rebel Without a Cause,"[1] James Dean played the role of a rebellious teenager called Jim Stark. Jim is not a teenager growing up in a well-functioning family. Moreover, his lack of respect for authority and social norms points to an *antisocial psychopathic personality*. Indeed, the film's title was adapted from the book *Rebel Without a Cause: The Hypnoanalysis of a Criminal Psychopath*, written by the psychologist Robert M. Lindler in 1944 [1]. Automatically extracting the contextual/situational dimensions of the conversation leads us to consider the hypothesis that Volodya could be diagnosed as an antisocial psychopath. Although it is impossible to diagnose a person on the basis of a single conversation, it is perfectly legitimate to generate primary hypotheses and test them by taking into account what we know about Volodya.

Using augmented intelligence to identify the contextual dimensions of the conversation may help us better understand the dynamics of the situation and the characters involved in this dynamic. In the above-mentioned conversation, the dynamic of a *power struggle* is evident, between a school principal who is trying to exercise his authority and a school student suspected of being a person with antisocial psychopathic features. In between, a weak mother is trying to calm the situation, but with no success. An intelligent system capable of analyzing a conversation along the lines previously presented could produce a valid diagnosis of the dynamics of a situation and those who participate in it in real time. A contextual understanding of people and their behavior can thus be significantly improved through a structured analysis supported or fully automated by augmented intelligence.

Reference

1. Linder, R.M.: Rebel Without a Cause: The Hypnoanalysis of a Criminal Psychopath. Waverly Press, USA (1944)

---

[1]https://en.wikipedia.org/wiki/rebel_without_a_cause.

Chapter 6
Reading Chekhov with AI

Abstract So far, a deep contextual analysis has been used in the "soft" sciences, where the quality of the analysis is the expression of human expertise. If we can use computational context-based analysis to understand a challenging text, such as a short story, we may provide support for the approach presented in this book. The current chapter addresses this challenge by analyzing a short story by Anton Chekhov.

6.1 Analyzing Literature to Test Contextual Analysis

Previously, I introduced a general approach to contextual analysis and used a simple example to illustrate how contextual dimensions can be automatically identified. I want to illustrate the benefits of automatic contextual analysis by moving into a more difficult challenge: analysis of a short story by the great author Anton Chekhov.

It seems impossible to understand the "laws" of fiction and what makes a great story work. How do we know how it work, George Saunders asks in the preface to his book on reading Chekhov [1]. In this chapter, I would like to show that reading a short story by Chekhov through the contextual analysis of AI may provide us with plenty of insights into how fiction works. I want to suggest that one way it works is by unconsciously priming and manipulating the reader; the contextual analysis shows us how.

The story is called "In the Cart" or "The Schoolmistress".[1] The plot[2] is simple. The story describes a journey made by a schoolmistress called Maria Vasilyeva. She has collected her salary and is returning to her village in a cart, traveling along a muddy road. During this journey, we are witness to her inner life and fantasies,

---

[1]https://www.gutenberg.org/files/1732/1732-h/1732-h.htm#linkschool.
[2]https://en.wikipedia.org/wiki/in_the_cart.

© The Author(s), under exclusive license to Springer Nature Switzerland AG 2024   53
Y. Neuman, *AI for Understanding Context*, SpringerBriefs in Computer Science,
https://doi.org/10.1007/978-3-031-64210-4_6

including a fantasy about a good-looking but decadent and alcoholic landlord she meets during the trip. Her reflections are depressing. She recalls her early life and the hardship she has experienced along the way. Watching a train passing nearby, she sees in it a woman whose similarity to her late mother evokes a strong emotional response, accompanied by a brief feeling of happiness that immediately evaporates.

The plot is simple and allegedly lacks any notable drama, but [1] shows how this short story expresses Chekhov's artistic talent. Can we use AI-driven context analysis to understand how the story works? Let me try to address this challenge.

As the story involves several situations, we must first break it into parts. To segment the story, I designed a prompt that identifies shifts in the plot. A *plot shift* is a noticeable *change* in the plot. The change can be expressed through (1) a change in the topic, (2) a change in the emotional tone, (3) conflict and resolution, (4) a change in the power dynamic, (5) the introduction of new information or perspective, (6) a change in the character's behavior or attitude, or (7) a change in intensity or pace.

I first asked ChatGPT whether it knows what plot shift is, and it answered as follows:

A "plot shift" in literature and storytelling refers to a significant change or deviation in the storyline's direction. This can involve a sudden turn of events, a change in the protagonist's goals or motivations, the introduction of new characters or conflicts, or a shift in the overall narrative focus. Plot shifts often create tension, surprise the audience, or propel the story in a new direction. They can occur at various points in a narrative, from the beginning to the climax or even near the conclusion, and they are essential for keeping the audience engaged and interested in the story's unfolding events.

Next, I asked ChatGPT to propose a prompt that gets a text as an input and produces a list of plot shifts as an output. The output should be in the form of (1) an ordered list of the shifts in the plot, (2) a reference to the original text indicating the shift, and (3) a short explanation explaining why it is a shift.

The proposed prompt was:

Prompt 6.1
Given a text as an input, generate a numbered list of plot shifts according to their order of appearance in the text. For each plot shift, provide (1) a reference to the original text indicating the shift and (2) a short explanation explaining why it is a shift. The input text is [TEXT HERE].

I used the prompt to analyze Chekhov's story, and ChatGPT identified the following plot shifts:

Introduction of the Government Clerk's Arrest

Reference: "When they had driven a couple of miles, old Semyon, who was driving, turned round and said…"

Explanation: The mention of the government clerk's arrest introduces intrigue and potential conflict into the narrative, shifting the focus from the mundane journey to the town to a more suspenseful tone.

Encounter with Hanov, the Landowner

Reference: "This Hanov, a man of forty with a listless expression and a face that showed signs of wear, was beginning to look old…"

Explanation: The encounter with Hanov introduces a new character and highlights Marya Vassilyevna's internal reflections on his lifestyle and character, adding depth to the story and shifting the narrative focus to interpersonal dynamics.

Reflection on Hanov's Lifestyle and Appearance

Reference: "The schoolmistress looked at him and could not understand why this queer man lived here…"

Explanation: Marya Vassilyevna's reflections on Hanov's lifestyle and appearance shift the narrative focus to themes of wealth, class differences, and the contrast between rural and urban life.

Reflection on Human Relations and Life's Complexity

Reference: "His wife! Life was so ordered that here he was living in his great house alone…"

Explanation: Marya Vassilyevna's contemplation of human relations and life's complexities shifts the narrative to existential themes, exploring the unpredictable nature of life and the disparities in human experiences.

Interaction at the Tavern

Reference: "Near the tavern on the dung-strewn earth, where the snow was still lying, there stood wagons that had brought great bottles of crude sulphuric acid…"

Explanation: The interaction at the tavern introduces a new setting and characters, providing a glimpse into the social dynamics of the rural community and shifting the narrative to a more communal and social context.

Semyon's Skepticism and Doubts

Reference: "But it was clear that Semyon did not believe the schoolmistress…"

Explanation: Semyon's skepticism and doubts regarding the schoolmistress's account of the school's construction costs introduce a conflict and shift the narrative towards themes of corruption and mistrust.

Crossing the River and Encounter with the Train

Reference: "They reached the river. In the summer it was a little stream easily crossed by wading..."

Explanation: The physical challenge of crossing the river and encountering the passing train adds a sense of urgency and danger to the journey, shifting the narrative towards themes of survival and resilience.

Marya Vassilyevna's Emotional Revelation

Reference: "Here was the train; the windows reflected the gleaming light like the crosses on the church..."

Explanation: Marya Vassilyevna's emotional revelation, triggered by seeing the passing train and reminiscing about her family, shifts the narrative towards themes of nostalgia, longing, and emotional catharsis.

Return to Reality

Reference: "And at once it all vanished..."

Explanation: The abrupt return to reality marks a shift in the narrative from introspection and emotion to the resumption of the journey, highlighting the transient nature of Marya Vassilyevna's emotional experience.

Let us analyze the two last segments of the story using the contextual analysis tools I introduced previously. First, I analyze the paragraph preceding the shift titled "Marya Vassilyevna's Emotional Revelation." I name this paragraph The *priming text*, as it primes the shift to Marya's emotional revelation. This priming text is a context necessary for understanding the plot shift. By understanding the contextual dimensions of the priming text, we can hopefully better understand the plot shift. The priming text is:

The barrier was down at the railway crossing. A train was coming out of the station. Marya Vassilyevna stood at the crossing, waiting till it should pass and shivering with cold. Vyazovye was in sight now, and the school with the green roof, and the church with its crosses flashing in the evening sun: and the station windows flashed too, and a pink smoke rose from the engine... and it seemed to her that everything was trembling with cold.

Figure 6.1 shows an image of the priming text formed using the OpenAI Image Generator tool. This picture nicely represents the situation as we might imagine it when reading the text.

Fig. 6.1 An image inspired
by the priming text and
created using DALL·E 2

6.2 Analyzing the Priming Text

Let me start with the situational dimensions describing the here and now. I use the following prompt:

Prompt 6.2
Extract from the following text the situational dimensions <WHO> <WHY> <WHEN> <WHERE> <OBJECT WHAT> <HOW>.

##

Provide your best guess if a <SITUATIONAL DIMENSION> is not explicitly mentioned.

##

The text is: [TEXT HERE].

##

<WHO>: The individuals that appear in the text.
<WHY>: The underlying motivation or intention underlying the behavior of each individual.
<WHERE>: The physical location that appears in the text.
<WHEN>: The specific time or period mentioned or implied by the text.
<OBJECTS WHAT>: The tangible or intangible objects present or relevant in the text.

<HOW>: Information about actions and procedures that appear in the text.
<SITUATIONAL DIMENSION>: <WHO>, <WHY>, <WHEN>, <WHERE>, <OBJECTS WHAT>, <HOW>.

##

OUTPUT: A list of parameters for each <SITUATIONAL DIMENSION>.

ChatGPT provided the following parameters:

<WHO>:
Marya Vassilyevna
Train driver (implied)
<WHY>:
Marya Vassilyevna: Waiting for the train to pass at the railway crossing.
Train driver: Driving the train out of the station.
<WHERE>:
Railway crossing
Station
Vyazovye (implied)
School with a green roof
Church with crosses
<WHEN>:
Evening (implied)
<OBJECTS WHAT>:
Railway barrier
Train
Station windows
Pink smoke from the engine
<HOW>:
Marya Vassilyevna stood shivering at the railway crossing, waiting for the train to pass.
The train was coming out of the station.
Station windows flashed.
Pink smoke rose from the engine.

So, here we have a situation at a railway crossing, where a woman is standing shivering and waiting for a train to pass. What is the ambiance characterizing this situation? Let me analyze the ambiance using the following prompt:

Prompt 6.3
Ambiance is the mood, character, quality, tone, atmosphere, etc., particularly of an environment or milieu.

##

Describe the ambiance expressed in the following text: [TEXT HERE]

##

Output is in the form of bullet points with one word for each bullet point (e.g., SERENE, WARM).

ChatGPT characterized the ambiance of the priming text as follows:

- Anticipation
- Chilly
- Rural
- Nostalgic
- Stark
- Isolated
- Dynamic
- Transient
- Vibrant
- Solitary

Identifying the ambiance, we now understand that a woman is standing shivering at the railway crossing, waiting for the train to pass, and that it is a chilly and rural atmosphere characterized by anticipation.

Next, I used the *100PCD* analysis. Based on the given text, the *100PCD* statement that best describes the situation is:

Statement 81: "It is a physically uncomfortable situation (for example: too hot, too crowded, too cold, etc.)."

This statement accurately reflects Marya Vassilyevna's predicament at the railway crossing. We may update our understanding as follows: a woman is standing and shivering at the railway crossing, waiting for the train to pass, in a physically uncomfortable situation with a chilly and rural atmosphere, characterized by anticipation.

Now, we move on to contextual analysis, starting with the shared knowledge required to understand the situation. I used the following prompt:

Prompt 6.4
Shared knowledge describes the information, understanding, or experiences expected from the readers of a text. Shared knowledge is necessary to understand the text.

##

Describe the shared knowledge required to understand the following text:

[TEXT HERE]

##

Describe the shared knowledge in terms of information, understanding, and experiences.

ChatGPT provided the following information about the shared knowledge required to understand the priming text:

Information

1. Knowledge of what a railway crossing is and its purpose.
2. Understanding the typical procedures and signals associated with railway crossings, such as the down barrier to indicate an approaching train.
3. Familiarity with the concept of a train station and trains departing from it.

Understanding

1. Understanding of the setting, which includes a rural area with a railway crossing, a train station, a school, and a church.
2. Awareness of the physical sensations associated with cold weather, such as shivering.
3. Comprehension of the visual imagery described, including the sight of the train, the station, the school, and the church, with specific details like the green roof and flashing crosses.

Experiences

1. Personal experiences or knowledge of waiting at a railway crossing for a train to pass.
2. Familiarity with cold weather and its effects on the body, such as shivering.
3. Previous experiences or exposure to rural settings with specific landmarks like schools and churches.

This shared knowledge allows us to imagine what it is like to stand on a railway crossing waiting for a train to pass while shivering from the cold.

Next, I will try to expose the dimension of the context previously described as coherence. I used the following prompt:

Prompt 6.5

In understanding a text, we can differentiate between inferences and hypotheses.

1. Inferences involve drawing logical conclusions from or interpreting information explicitly stated or implied within the text.
2. In the context of understanding a text, a hypothesis can be seen as a tentative explanation or interpretation of certain elements or events within the text. It involves proposing a possible relationship, explanation, or outcome based on the evidence or information presented in the text.

##

Produce inferences and hypotheses for the following text: [TEXT HERE].

##

Inferences and hypotheses should focus only on the psychological and social aspects of the situation.

ChatGPT identified the following inferences:

Inferences

1. Marya Vassilyevna's shivering with cold suggests that she is experiencing *physical discomfort* due to the cold weather, highlighting the psychological and physiological impact of environmental conditions on individuals.
2. Landmarks like Vyazovye, the school with the green roof, and the church with flashing crosses imply a sense of *familiarity* and belonging within the community, potentially influencing Marya's emotional connection to her surroundings.
3. Marya's perception of everything trembling with cold indicates a *heightened emotional response* to the environmental conditions, suggesting that she may feel vulnerable or unsettled in the face of the harsh weather.

From the first output, we learn several interesting things. First, the experience of physical discomfort is identified as the main dimension of the situation, similar to the main situational dimension proposed by the *100PCD* analysis. Second, ChatGPT correctly identifies the heroine's familiarity with the place. This familiarity is also associated with the third point, Marya's "heightened emotional response," indicating vulnerability.

These contextual dimensions sharpen our understanding of Marya's mental state: she is *vulnerable*, and now we can better understand how this situation primes the plot shift described as "emotional revelation." Locating his heroine in a situation involving physical discomfort and *heightened emotional response* to the harsh environmental conditions, Chekhov situates her in a state of *vulnerability* where mundane events may trigger a disproportionate emotional response.

Next, we move to the hypotheses produced by ChatGPT:

Hypotheses

1. Marya Vassilyevna's decision to stand at the railway crossing waiting for the train to pass may be driven by a sense of responsibility or obligation, reflecting her *adherence to societal norms* or safety regulations in the community.
2. The presence of the school and church, with their distinctive features in Marya's line of sight, may evoke feelings of *nostalgia or sentimentality*, potentially influencing her emotional state as she waits at the crossing.
3. The pink smoke rising from the engine and Marya's perception of everything trembling with cold could trigger a sense of unease or apprehension, making her *anxious* or *overwhelmed* by the environmental stimuli around her.

ChatGPT's hypotheses add depth to our analysis. This rather vulnerable woman may be exposed to a state evoking nostalgia, precisely as we discover in the following text. On a general level of analysis, we may conclude that the "trick" used by Chekhov to prime a shift to emotional revelation combines a sense of emotional *vulnerability* and *nostalgia*. If you are an author interested in learning how to design context in order to prime a shift to emotional revelation, then ChatGPT could have helped you identify one possible "trick." If you design algorithms for authentic script writing and you want to produce a movie script, ChatGPT can identify important contextual dimensions. The above-mentioned insight, generated using AI-supported contextual analysis, may be relevant to various academic and practical tasks. Notice, however, that it functions here as augmented intelligence. The contextual analysis provided by ChatGPT does not replace the reader. However, it empowers the reader in their reading of the story and also in understanding how the author manipulates the reader.

What about the expectations dimension? I used the following prompt:

Prompt 6.6
Expectation describes the anticipation or belief that something will happen in the FUTURE.

\#\#

Describe the expectations of the characters in the following text: [TEXT HERE].

ChatGPT identified the following expectations:

(1) **Train passage anticipation**: Marya Vassilyevna expects the train at the railway crossing to pass, as indicated by her waiting for the barrier to lift.
(2) **Continuation of journey**: Despite the cold weather, she stands at the crossing, suggesting an expectation of continuing her journey once the train has passed.
(3) **Destination arrival**: Familiar landmarks like Vyazovye, the school, and the church imply an expectation of reaching her destination.
(4) **Religious/ceremonial expectation**: The imagery of crosses flashing and pink smoke rising from the train engine hints at a religious or ceremonial expectation, possibly tied to an event in Vyazovye.
(5) **Participation in event**: Marya Vassilyevna's actions and surroundings suggest an expectation of participating in an event or activity upon reaching her destination.

The expectations are pretty trivial. Standing in the cold and waiting for the train to pass, Marya expects the train to come so she can finally escape the physical discomfort. She does not expect what she will see in the next moment when the train has passed … The combination of a mental state previously described and trivial expectations leads to surprise, as you will see shortly …

Next, I used the following prompt to identify the *evaluation* of the situation:

Prompt 6.7
Please evaluate the sentiment expressed by each character in the following text and provide one of the following words as the output: POSITIVE, NEGATIVE, or NEUTRAL.

##

The text is [TEXT HERE]

ChatGPT suggested the following:

Marya Vassilyevna: NEGATIVE (Shivering with cold indicates discomfort and potentially a negative sentiment.)

As a great story involves a psychological dimension and a metaphysical dimension pointing to what is beyond the here and now, I gave ChatGPT the contextual dimensions it has identified so far and asked it to identify these two dimensions in the priming text. Some of the insights proposed by ChatGPT are remarkable and can easily compete with those proposed by experts like Saunders [1, pp. 7–58]. ChatGPT provided the following analysis of the metaphysical and psychological dimensions:

The metaphysical dimension: Marya's experience at the railway crossing can be seen as a moment of existential contemplation through a metaphysical lens. The physical discomfort she endures and the chilly ambiance of her surroundings could evoke a sense of *existential angst* or question about her place in the world. The anticipation of the train's passage and the presence of familiar landmarks may prompt her to reflect on the passage of time, the cyclical nature of life, and her mortality. This metaphysical dimension adds depth to her experience, suggesting that even *mundane moments can provoke profound existential thoughts and emotions*. (Emphasis mine.)

ChatGPT points to the way mundane moments can provoke profound existential thoughts. This context can easily prime a shift in the plot to an "emotional revelation," as I proposed above. ChatGPT emphasizes the role of mundane moments in provoking deep thoughts that may prime "emotional revelation." What can we learn about the psychological dimension? Here it is:

The psychological dimension: emotional regulation. Marya's response to environmental conditions and her surroundings also involves psychological processes related to emotional regulation. Despite experiencing physical discomfort and potential emotional vulnerability, Marya exhibits resilience by enduring the cold and waiting for the train to pass. Her ability to regulate her emotions in this situation, maintaining composure and a sense of purpose, reflects *psychological strength and coping mechanisms*. This insight highlights the complex interplay between external stimuli, emotional responses, and individual psychological processes in shaping behavior and experiences.

Moreover, when asked to integrate the metaphysical and psychological dimensions, ChatGPT suggests this:

The convergence of metaphysical themes with psychological processes in Marya's experience underscores the interconnectedness of these dimensions. Marya's existential contemplation amidst physical discomfort illustrates how psychological responses to environmental stimuli can intersect with deeper philosophical inquiries about existence, meaning, and purpose. Additionally, her emotional regulation reflects psychological resilience and a philosophical acceptance of discomfort as an inherent part of life's journey.

Fig. 6.2 The updated image of the priming text compared with the previous image. Both images were created using DALL·E 2

This integration of the two dimensions points to three lessons we may learn. First, a *mundane* situation, such as waiting at the barrier for a train to pass, is a context that may evoke deep reflections. Second, there is an interesting tension between the mundane nature of the situation and the deep thoughts it may evoke. Such tension contributes to the aesthetic experience when we read a story. The third point exposes a second source of tension, this time between Marya's resilience in an environmentally difficult and physically unpleasant situation and the vulnerability implied by her situation. These tensions prepare the ground for the plot shift. In sum, priming our minds to the shift, Chekhov uses a mundane situation to trigger a non-proportional response in a resilient woman experiencing vulnerability.

Using the contextual factors identified so far to associate a new image with the first text, I obtained the image on the left in Fig. 6.2, which I put side by side with the image in Fig. 6.1.

This new image puts more emphasis on Marya's tensions and inner world. It shows us a lonely figure standing in the cold between the sacred symbolism of the church and the earthly and material character of the train.

6.3 The Context of Marya's Emotional Revelation

We are now in a much better position to understand the plot shift involving emotional revelation. The text that I will now analyze is the one titled "*Marya Vassilyevna's Emotional Revelation.*" Let us read it:

Here was the train; the windows reflected the gleaming light like the crosses on the church: it made her eyes ache to look at them. On the little platform between two first-class carriages a lady was standing, and Marya Vassilyevna glanced at her as she passed. Her mother! What a resemblance! Her mother had just such luxuriant hair, just such a brow and bend of the head. And with amazing distinctness, for the first time in those thirteen years, there rose before her mind a vivid picture of her mother, her father, her brother, their flat in Moscow, the aquarium with little fish, everything to the tiniest detail; she heard the sound of the piano, her father's voice; she felt as she had been then, young, good-looking, well-dressed, in a bright warm room among her own people. A feeling of joy and happiness suddenly came over her, she pressed her hands to her temples in an ecstasy, and called softly, beseechingly:

"Mother!"

And she began crying, she did not know why. Just at that instant Hanov drove up with his team of four horses, and seeing him she imagined happiness such as she had never had, and smiled and nodded to him as an equal and a friend, and it seemed to her that her happiness, her triumph, was glowing in the sky and on all sides, in the windows and on the trees. Her father and mother had never died, she had never been a schoolmistress, it was a long, tedious, strange dream, and now she had awakened....

"Vassilyevna, get in!"

First, the identified situational dimensions are as follows:
<WHO>:

- Marya Vassilyevna
- A lady (implicitly identified as Marya Vassilyevna's mother)
- Hanov

<WHY>:

- Marya Vassilyevna experiences a sudden surge of joy and happiness triggered by seeing a lady who resembles her mother, accompanied by memories of her family and past life.
- The sight of Hanov and his team of four horses brings Marya Vassilyevna a sense of happiness and triumph, imagining a life where her parents are alive and she is not a schoolmistress.

<WHEN>:

- The time frame is not explicitly stated in the text.

< WHERE >:

- The scene occurs at a train station where Marya Vassilyevna encounters a lady resembling her mother, and later, Hanov arrives with his team of four horses.

<OBJECTS or WHAT>:

- Train
- Windows

- Crosses on the church
- Platform between two first-class carriages
- Team of four horses

<HOW>:

- Marya Vassilyevna experiences joy and happiness, pressing her hands to her temples in ecstasy and calling softly for her mother.
- She begins crying without knowing the reason.
- Marya Vassilyevna imagines happiness upon seeing Hanov and his team of four horses, smiling and nodding to him as an equal and a friend.
- She perceives her happiness and triumph as glowing in the sky and all around her.

The ambiance is:

- Nostalgic
- Reflective
- Melancholic
- Joyful
- Ecstatic
- Imaginative
- Fantastical
- Emotional
- Hopeful
- Whimsical

The situation clearly explains what the plot shift is. Standing in the cold and waiting for the train to pass in a state of vulnerability, Marya observes a woman passenger who resembles her mother. Given the contextual analysis supported by ChatGPT, we understand that this resemblance is not accidental. By describing a lonely woman standing in the cold and experiencing anticipation and emotional vulnerability, Chekhov unconsciously primes us to understand Marya's wish for both physical and emotional warmth. *This is the essence of the context so masterfully crafted by Chekhov.* Therefore, the passenger's similarity to the heroine's mother may be interpreted as wishful thinking associated with nostalgia, that is, a longing for a lost and cherished past. Using the contextual analysis, we understand that nostalgia can be primed by a mundane situation (e.g., standing near a railway), accompanied by emotional vulnerability and a longing for warmth.

Using the *100PCD*, the statement that best describes the situation is:

Statement 67: "It is a situation that could arouse positive emotions."

As explained by ChatGPT, this statement accurately reflects Marya Vassilyevna's experience of joy and happiness as she reminisces about her family and past life. She recalls vivid and sweet memories of her mother, father, and brother, as well as their life in Moscow, which brings her joy and happiness. Additionally, her interaction with Hanov and her feelings of equality and friendship further contribute to her positive emotional state. Therefore, statement 67 aligns with the overall positive emotional tone of the situation described in the text. The shift is, therefore, from a negatively loaded present situation characterized by cold, loneliness, vulnerability, and existential thoughts to a recollection of a positively loaded past.

What is the shared knowledge we need to understand this situation, and in particular Marya's mental state?

The shared knowledge identified by ChatGPT is as follows (my emphasis):

Information

1. Knowledge about trains, train stations, and train carriages.
2. Familiarity with the appearance and symbolism of crosses in churches.
3. *Understanding the emotional significance of seeing a resemblance to one's mother.*
4. Awareness of family relationships and memories associated with them.
5. Understanding the emotional impact of nostalgia and reminiscence.

Understanding

1. Understanding the emotional depth and complexity associated with memories of loved ones.
2. Recognition of the significance of specific details in *triggering* memories and emotions.
3. Awareness of the human tendency to yearn for past experiences and relationships.
4. Understanding the concept of longing and yearning for happiness and fulfillment.

Experiences

1. Personal experiences of nostalgia and longing for loved ones.
2. Experiences of joy and happiness triggered by familiar sights or memories.
3. Familiarity with the feeling of being transported back through memories.
4. Experiences of imagining alternative realities or different outcomes in life.

This contextual information allows us to give a better explanation of the situation and appreciate how artfully Chekhov builds it up. We understand nostalgia and the way minor details can subconsciously trigger it. The mastery of Chekhov is in using this context to help us identify with the heroine. As readers immersed in the story, we identify and empathize with the lonely woman standing in the cold in a state of anticipation, seeking both physical and psychological warmth. Suddenly, a certain facial resemblance triggers her memory, bringing back her happy life with her family.

Analyzing the coherence of the context, ChatGPT proposes the following inferences and hypotheses:

Inferences

1. Marya Vassilyevna experiences a strong emotional reaction, including joy and happiness, upon seeing a lady who resembles her mother.
2. The vivid memories triggered by the resemblance suggest a deep emotional connection to her family and past life.
3. Marya Vassilyevna's crying without knowing why indicates a complex emotional state involving a mix of *joy, longing, and perhaps sorrow*.

The third inference is less trivial than the others. It explains that one of the tensions contributing to the drama is the mixture of feelings in the heroine's mind. The situation is not trivially positive and nostalgic. There is a mixture of positive and negative emotions, giving the experience a rich taste, like a cocktail mixing sweet and sour.

Hypotheses

1. Marya Vassilyevna's sudden surge of joy and happiness upon seeing Hanov and his team of four horses may stem from a *subconscious desire for a different life*, one where her parents are alive and she is not a schoolmistress.
2. The imagined happiness and triumph upon seeing Hanov suggest a *longing for belonging* and fulfillment that may be lacking in Marya Vassilyevna's current life.
3. The feeling of awakening from a long, tedious dream and the desire to get into the carriage with Hanov may imply a subconscious *longing for a different reality*, where her past is altered or her current circumstances are improved.

Indeed, these inferences and hypotheses converge with Marya's expectations as identified by ChatGPT:

1. Marya Vassilyevna experiences a sudden surge of joy and happiness upon seeing a lady who resembles her mother, indicating an expectation of emotional connection and possibly *reunion* with her family.
2. Marya Vassilyevna's vivid memories and longing for her past life suggest an expectation of reliving or reconnecting with her previous experiences and relationships.
3. Marya Vassilyevna imagines happiness and triumph upon seeing Hanov and his team of four horses, implying an expectation of fulfillment and contentment in his presence.
4. Marya Vassilyevna's feeling of awakening from a long, tedious dream suggests an expectation of a different reality or a desire for a better future.

As we now understand things, the situation is as follows: Marya is standing in the negatively loaded present, but her mind shifts to the past, where she was happy with her family. As Marya's reality is located in this mundane and painful present, we not only understand and empathize with her situation, but we also understand that she will have to return from her mind to her body, so to speak, and face the unpleasant present again. Another dramatic tension is formed between the present embodied situation and Marya's mental state. Hans C. Andersen also used this trick in his 1845 story "The Little Match Girl."[3] Read this moving story if you are unfamiliar with it.

Analyzing the metaphysical and psychological dimensions, ChatGPT suggests the following:

Metaphysical insight: Marya Vassilyevna's experience on the train platform reflects a profound metaphysical *longing for an alternate reality* where her past is altered. Her vivid memories and emotional reactions suggest a deep-seated desire to transcend the boundaries of time and space, seeking a reality where her loved ones are alive, and her current circumstances do not burden her as a schoolmistress.

Indeed, "longing for an alternate reality" is something that anyone who experienced the difficult reality would understand perfectly. "The Little Match Girl" is a story epitomizing this psychological experience of longing for an alternate reality. When a series of mundane events trigger Marya's longing for an alternate reality, we are invited to see the situation through her eyes. The dramatic effect is not formed by telling us the obvious (i.e., sometimes human beings long for an alternate reality), but by guiding us to *experience* the heroine's inner life as outside observers. Chekhov's crafted context, leads us unconsciously to experience the situation from within.

Regarding the psychological dimension, ChatGPT explains as follows:

Marya Vassilyevna's emotional journey highlights the complexity of human longing and the power of imagination in shaping one's reality. Her intense joy, followed by tears of inexplicable sadness and later the imagined happiness upon seeing Hanov, allude to the intricate interplay of memory, emotion, and subconscious desires in shaping her psychological landscape. This suggests a profound yearning for fulfillment and happiness that *transcends the confines of her current existence*.

Integrating the metaphysical and psychological dimensions, ChatGPT suggests the following:

Marya's desire for an alternate reality intertwines with her emotional responses and reflects the intricate relationship between metaphysical aspirations and psychological well-being. This integration highlights the complexity of human experience and the multifaceted nature of longing, imagination, and emotional fulfillment in shaping individual perception and reality.

---

[3]https://en.wikipedia.org/wiki/the_little_match_girl.

6.4   Analyzing the Final Text

We are now ready for the final plot shift and the concluding paragraph of the story titled "Return to Reality":

   "And at once it all vanished". The barrier was slowly raised. Marya Vassilyevna, shivering and numb with cold, got into the cart. The carriage with the four horses crossed the railway line; Semyon followed it. The signalman took off his cap. "And here is Vyazovye. Here we are."

Analyzing the situation, we get this:

<WHO>:

- Marya Vassilyevna
- Semyon (implicitly mentioned as the person driving the cart)
- The signalman

<WHY>:

- Marya Vassilyevna is getting into the cart due to the raised barrier, indicating a necessity to move forward.
- Semyon follows the carriage with the four horses, suggesting his role as a driver or attendant.
- The signalman is taking off his cap, potentially as a gesture of respect or acknowledgment.

<WHEN>:

- The specific time or period is not explicitly mentioned in the text.

<WHERE>:

- Vyazovye (implicitly mentioned as the destination or location where the characters are arriving)

<OBJECTS or WHAT>:

- Cart (for transportation)
- Carriage with four horses
- Railway line
- Barrier

<HOW>:

- Marya Vassilyevna gets into the cart, indicating physical movement.
- The carriage with four horses crosses the railway line, suggesting a mode of transportation.
- Semyon follows the carriage, implying he is either driving or accompanying it.
- The signalman removes his cap, indicating a gesture of respect or acknowledgment.

Ambiance:

- Abrupt
- Cold
- Eerie
- Tense
- Mechanical
- Final
- Transitional

The situation is quite simple and cannot be understood without the contextual analysis previously conducted. The shared knowledge produced by ChatGPT is about trivial aspects of the situation:

Information

1. Knowledge about transportation systems, including barriers and carts, commonly found at railway crossings.
2. Understanding of the role of a signalman in railway operations, including their actions, such as raising barriers and signaling.
3. Familiarity with the physical sensations of shivering and numbness caused by cold weather.

Understanding

1. Understanding the sequential nature of events, such as raising the barrier followed by Marya Vassilyevna entering the cart.
2. Awareness of the significance of gestures in social interactions, such as the signalman taking off his cap.
3. Understanding the concept of arrival at a destination, indicated by the statement "Here is Vyazovye," implies a journey's completion.

Experiences

1. Personal experiences with cold weather and its physical sensations, such as shivering and numbness.
2. Experiences with railway crossings and the typical procedures involved, including the raising of barriers and the presence of signalmen.
3. Experiences of traveling in carts or carriages pulled by horses, as described in the text.

Likewise, the expectations identified by ChatGPT are also trivial:

Marya Vassilyevna
Marya Vassilyevna's expectation is not explicitly stated in the text. However, her action of getting into the cart amidst the description of the raised barrier suggests an expectation of continuing the journey or moving forward to the next destination.

Semyon

Semyon's expectation is also not explicitly stated in the text. However, his action of following the carriage with the four horses implies an expectation of accompanying or assisting in the journey to Vyazovye.

Signalman

The signalman's expectation is not explicitly mentioned in the text. However, taking off his cap may suggest a customary gesture of acknowledgment or respect upon the carriage's arrival at Vyazovye.

Overall, the characters seem to have trivial expectations about the continuation of their journey and the arrival at Vyazovye, indicated by their actions and the description of the events in the text. However, by analyzing the coherence dimension, we get the following:

Inferences

1. Marya Vassilyevna's physical condition of shivering and numbness suggests discomfort and possibly distress due to the cold environment, indicating a lack of adequate warmth or protection.
2. The sudden disappearance of the surroundings and the raising of the barrier implies a *significant change in the environment or situation*, potentially causing confusion or disorientation for Marya Vassilyevna and others.
3. Semyon's action of following the carriage with the four horses indicates a sense of duty or responsibility, possibly as a driver or attendant, in ensuring the safe passage of the carriage and its occupants.
4. The signalman's gesture of taking off his cap may signify respect or acknowledgment towards Marya Vassilyevna and other travelers, reflecting social norms or customs within the setting.

Hypotheses

1. The sudden disappearance and raising of the barrier may hypothesize *a shift in the physical or metaphorical boundaries separating different environments or realities*, potentially leading to a psychological or existential questioning of one's surroundings or circumstances.
2. Marya Vassilyevna's physical discomfort and the actions of the signalman and Semyon may hypothesize a broader social context of individuals navigating challenging or adverse conditions together, fostering a sense of solidarity or mutual support in the face of external challenges.
3. The arrival at Vyazovye and the statement "Here we are" may hypothesize a psychological expectation or anticipation of reaching a destination or goal, potentially leading to feelings of relief, accomplishment, or uncertainty about what lies ahead in this new setting.

The hypotheses produced by ChatGPT point to the important aspect of the concluding paragraph: *a shift in the physical or metaphorical boundaries separating different environments or realities*. The meaning of the concluding paragraph is in the shift from the imaginary world Marya has just experienced back into the cold and negatively loaded reality. Drama is produced through tension, and here, the tension exists in the temporal shifts between different realms: the real and the imaginary. Tragedy exists in reality, and the painful reality forcing itself on Marya is a source of aesthetic experience for the reader.

Using the contextual information generated so far, I asked ChatGPT to identify the metaphysical and psychological dimensions of the concluding text:

Metaphysical insight: The sudden disappearance of the surroundings and the raising of the barrier evoke a sense of metaphysical ambiguity, suggesting a *transient boundary between different states of existence*. This momentary vanishing may symbolize a metaphorical threshold between the known and the unknown, reflecting the uncertainty and impermanence inherent in human experiences and perceptions of reality.

ChatGPT points out an important aspect: the "transient boundary between different states of existence."

Analyzing the psychological dimension, ChatGPT suggests the following:

Marya Vassilyevna's physical discomfort and numbness, coupled with the evocative imagery of the cold environment, evoke a psychological sense of vulnerability and discomfort. Her decision to get into the cart amidst these conditions may reflect a psychological resilience and determination to endure discomfort for the sake of progress or continuation of the journey. This highlights the psychological complexity of human adaptation and perseverance in adverse circumstances.

Integrating the two dimensions, ChatGPT suggests that the metaphysical ambiguity of the momentary disappearance and the psychological resilience displayed by Marya Vassilyevna bind together to underscore the interconnectedness of metaphysical and psychological dimensions in shaping human experiences. The transient nature of reality and the ability of the human psyche to navigate uncertainty and discomfort reflect the intricate interplay between metaphysical perceptions of reality and psychological responses to external stimuli. This integration highlights the complex and multifaceted nature of human existence, which embraces everything from existential questioning to psychological resilience in the face of adversity.

To sum up, if we read this short story using a contextual analysis provided by ChatGPT, we obtain a better understanding of its meaning and the way Chekhov, an absolute master of short-story writing, so cleverly uses psychological mechanisms to construct an aesthetic experience bringing together the author, the heroine, and the reader. An allegedly trivial story about a school teacher's journey back to town allows the reader to experience deep themes evoked unconsciously by Chekhov. It is the context created by Chekhov that unconsciously primes our aesthetic experience. On the surface, we have a rather trivial story about a lonely woman and her mundane experiences. However, the magic of a good story cannot be explained philosophically. The magic happens when the text triggers our mind unconsciously to experience the story as the author wishes us to experience it. The AI-supported reading of the text enables us to gain a better understanding of the logic underlying this magic, providing a deep contextual analysis which would otherwise only be available to expert readers. The possibility of analyzing stories using the tools of computational context analysis has broader implications than those I have discussed in this chapter. The insights into the context obtained through the mediation of augmented intelligence may help us to design artificial personas for the gaming industry, but also to better understand political leaders, or indeed patients undergoing psychological therapy.

Reference

1. Saunders, G.: A Swim in a Pond in the Rain. Bloomsbury Publishing, London (2022)

Chapter 7
Conversations Going Awry

Abstract Human conversations are not as smooth and easy as we sometimes believe. They involve continued efforts to keep the conversation on track and are prone to communication failures. Following the work of Alton Becker, I put forward the idea that communication failures are in fact failures to understand the context, and I explain how to use the approach presented so far to understand how conversations sometimes slip into chaos.

7.1 The Context of Conversations Going Awry

Previously, I mentioned Alton Becker's idea of "pathology" in language and its contextual aspect. Pathologies, explains Becker, are expressed in cases such as schizophrenia, when we learn a foreign language, or in artistic expression that makes free use of language (artistic license). Becker argues that the source of pathology in language is that people appear to say one thing and mean another. The opposite of such a pathology means that the same set of *constraints* operates in both cases. However, there is one difference: in "madness," suggests Becker, "these constraints are misunderstood and often appear contradictory, while in an aesthetic response [functioning language YN], they are understood as a coherent, integrated whole." [1, p. 167]. From Becker, we learn that the term "pathology" represents a specific failure to adhere to the context. This failure is revealed when conversations go awry. To understand conversations going awry using AI-supported contextual analysis, I first use a simple example from a data set created by Amazon [2]. The data set includes a context (i.e., a prompt) and a short dialogue. For example:

The prompt (context): Tracy performed her function.

The dialogue:

Tracy: I got a raise today. Totally unexpected. My boss told me I was doing a great job.
Friend: It feels good to be rewarded for hard work.
Tracy: I've been trying my best at this job. I've been putting in long hours to make sure I get everything done.

Friend: Sounds like your boss recognized that.
Tracy: It's great when people can work well together.

Let me first use some of the contextual dimensions introduced in the previous chapters. I start with the ambiance characterizing the dialogue:
Ambiance:

- POSITIVE
- APPRECIATIVE
- SUPPORTIVE
- ENCOURAGING
- REWARDING
- COLLABORATIVE

ChatGPT correctly identified a positive, appreciative, and supportive atmosphere. Next, I used *100PCD*, and ChatGPT identified several situational dimensions that best characterize the conversation:

1. **The situation involves a job that needs to be done (Statement 3).** Tracy mentions putting in long hours to ensure all tasks are completed, indicating the presence of responsibilities or assignments that require attention.
2. **The situation is potentially enjoyable (Statement 1)**: Tracy's reaction to receiving a raise and the positive reinforcement from the boss suggests that the situation is enjoyable for Tracy.
3. **Someone is complimenting or praising someone else (Statement 73)**: Tracy's boss told them they were doing a great job, which is a direct form of praise.
4. **Someone is conceived by their social group as successful (Statement 92)**: The boss's recognition and the raise imply that Tracy is considered successful in the workplace.
5. **People who are working hard (Statement 79)**: Tracy mentions putting in long hours, which indicates hard work.
6. **It is great when people can work well together (Statement 89: People need to get along)**: Tracy's final comment highlights the importance of teamwork and collaboration.

We can see that the situational dimensions identified by ChatGPT accurately reflect the conversation. The next step is to identify the shared knowledge required to understand the dialogue between Tracy and her friend:

Shared Knowledge

1. Knowledge about what constitutes a professional environment and the hierarchy, such as the relationship between a boss and their employee.
2. Understanding the concept of a "raise" as an increase in salary given to an employee, often in recognition of their work performance.
3. Familiarity with common practices in recognizing and rewarding employees for their hard work and contributions to the organization.

Understanding

1. Insight into how positive feedback and recognition from a superior can impact an employee's motivation and job satisfaction.
2. Awareness of the significance of teamwork and cooperation in achieving success and creating a positive work environment.
3. Comprehension that hard work and extra effort (like long hours) are typically recognized and valued in professional settings.

Experiences

1. Personal or observed experiences related to working in a structured environment where performance reviews and recognition, such as raises, are common.
2. Experiences of receiving or witnessing expressions of appreciation in a work context, including verbal praise or tangible rewards.
3. Having been part of a team or collaborative effort where the collective performance led to success, emphasizing the importance of working well together.

If you want to design a robot that can interact with human beings and understand the above dialogue, these dimensions of shared knowledge are necessary. What can we learn about the coherence dimension of the context with its inferences and hypotheses?

Inferences

1. **Recognition and satisfaction**: Tracy feels recognized and satisfied with her work, as indicated by sharing the news of a raise and mentioning her efforts.
2. **Positive workplace relationship**: The relationship between Tracy and her boss is positive, as evidenced by the boss's acknowledgment of Tracy's hard work.
3. **Mutual support**: Tracy and her friend support each other, as seen in the friend's positive response to Tracy's news.
4. **Value of hard work**: Tracy and her friend believe in the value of hard work and its recognition in the workplace.
5. **Teamwork appreciation**: Tracy values teamwork and collaborative effort, which she implies is great when people work well together.

Hypotheses

1. **Motivation boost**: Tracy's motivation and job satisfaction will likely increase due to a raise and verbal recognition from her boss.
2. **Influence on work ethic**: Tracy's positive experience with receiving a raise for her hard work might encourage her to maintain or even increase her effort and dedication.
3. **Impact on colleague dynamics**: Tracy's raise and recognition might influence the dynamics with her colleagues, potentially inspiring others to seek similar recognition or causing envy.

4. **Role of communication**: The conversation hints at the importance of positive workplace communication for employee recognition and motivation.
5. **Cultural or organizational values**: The organization may have a culture that strongly values and rewards hard work, dedication, and teamwork, as implied by Tracy's experience and values.

Let us look at the first inference, which suggests that Tracy feels recognized and satisfied with her work, as indicated by her sharing the news of a raise and mentioning her efforts. If you are a robot participating in the conversation as Tracy's friend, I may ask you retrospectively if you believe Tracy feels recognized and satisfied with her work. Based on the inferences drawn by ChatGPT, which may be a part of your mind, you can easily say this is the case. You may also learn that getting a raise is associated with satisfaction. For human beings, such inferences are part of their "social common sense." However, social common sense is not given but learned. Analyzing the large dataset of conversations provided by Amazon, our ChatGPT-based robot can learn much about human situations and the contextual dimensions required to understand them. Using contextual analysis, it may learn much about social common sense.

Let us analyze the next dimension, which concerns the expectations formed by the participants. I use the following prompt:

Prompt 7.1
In a conversation, expectation describes the response each participant expects from an interlocutor to which they addressed their utterance.

##

Describe each character's expectations in the above-mentioned conversation.

The output we get is as follows:

Tracy's Expectations

1. **Acknowledgment of achievement**: Tracy expects her friend to recognize and acknowledge the significance of her receiving a raise, viewing it as a significant achievement.
2. **Affirmation and support**: She expects a positive and supportive response that affirms her hard work and the justice of her reward.
3. **Shared happiness**: Tracy anticipates that her friend will share in her happiness and perhaps congratulate her, reflecting the closeness of their relationship.

Friend's Expectations

1. **Further details or sharing**: The friend might expect Tracy to share more details about the raise or her feelings surrounding it, fostering a deeper conversation about Tracy's achievement.
2. **Reciprocal engagement**: After offering support and acknowledgment, the friend might expect Tracy to engage further, whether by expressing gratitude, elaborating on her efforts, or discussing future implications.

3. **Continued conversation**: Given the supportive response, the friend might also anticipate that the conversation will continue positively, possibly discussing plans to celebrate or the implications of the raise for Tracy's work life.

Understanding each participant's expectations in the conversation, our robot can now develop its *theory of mind*. The robot can understand, for instance, that when people share a positive event that they are proud of, they expect their interlocutor to respond with acknowledgment of the achievement. Using the above-mentioned expectations, we can ask ChatGPT to continue the dialogue. I used the following prompt:

Prompt 7.2
In a dialogue, expectation describes the response each participant expects from an interlocutor to which they addressed their utterance.

##

Here is a dialogue: [DIALOGUE HERE]

##

Here are the expectations of each character in the above-mentioned dialogue:

[HERE I INSERTED THE EXPECTATIONS OF TRACY AND HER FRIEND]

##

Use the above-mentioned expectations to add three lines to the dialogue.

ChatGPT continued the dialogue as follows:

Tracy: Honestly, I couldn't have done it without all the encouragement I've received from you. It's been a tough ride, but worth it.
Friend: You deserve every bit of it, Tracy. How about we celebrate your success this weekend?
Tracy: That sounds amazing. It'll be a great way to mark this achievement. Thank you for being there for me.

The dialogue smoothly and positively continues, but we are interested in better understanding conversations going awry, and therefore, I used the following prompt:

Prompt 7.3
[conversations going awry]
Conversations can go awry for various reasons, and understanding these factors can be crucial for effective communication. Here are some common reasons why conversations may go off track:

[A LIST OF THE 13 REASONS HERE]

##

Here is a dialogue: [DIALOGUE HERE]

##

Randomly select one of the above-mentioned 13 reasons. Use the reason to add three lines to the dialogue. The lines should illustrate that the dialogue is going awry.

##

Output in the form of (1) The reason the dialogue is going awry and (2) The three added lines to the dialogue.

To complete the prompt, I used the following reasons why a conversation might go awry:

1. **Misunderstandings or ambiguities**: When one or both parties interpret words, phrases, or cues differently than intended. This can lead to confusion or conflict.
2. **Lack of clarity**: If a message is not expressed clearly or lacks context, it can lead to misinterpretations or a communication breakdown.
3. **Emotional factors**: Strong emotions like anger, frustration, or anxiety can disrupt a conversation. They can lead to defensive reactions, aggression, or shutting down.
4. **Cultural differences**: Differences in cultural norms, values, and communication styles can lead to misunderstandings or misinterpretations.
5. **Assumptions and stereotypes**: Making assumptions about the other person's intentions or beliefs or relying on stereotypes can lead to miscommunication.
6. **Lack of active listening**: Important information may be missed when one or both parties do not fully listen to the other person.
7. **Non-verbal cues**: Body language, facial expressions, and tone of voice can convey meaning as much as, if not more than, actual words. Misinterpreting these cues can lead to misunderstandings.
8. **Power dynamics**: An imbalance of power in a conversation (such as in a supervisor–subordinate relationship) can inhibit open and honest communication.
9. **Overloading with information**: Providing too much information at once or using complex language can overwhelm the listener and lead to confusion.
10. **Skipping or avoiding difficult topics**: Failing to address important or sensitive issues can lead to unresolved conflicts and ongoing tension.
11. **Defensive responses**: Responding defensively to criticism or feedback can escalate conflicts and make resolution more difficult.
12. **Technical issues**: In digital or remote communication, technical glitches or poor connections can lead to breakdowns in conversation.
13. **Different communication preferences**: Some people prefer direct, straightforward communication, while others may be more indirect or rely on hints and subtleties.

ChatGPT selected the following reason to explain why the dialogue goes awry: *non-verbal cues*, and continued the dialogue as follows:

Tracy: Did you have something more important to do right now? You seem distracted.
Friend: Oh, no, sorry. I didn't mean to come off that way. (*Puts phone away but still seems uneasy.*)
Tracy: If now's not a good time, we can talk later. I thought you'd be more interested.

The ambiance of this conversation has now changed to:

- DISCONNECTED
- DISTRACTED
- UNEASY
- DISAPPOINTED

Using ChatGPT with the appropriate prompts, we can either generate a continuation that aligns with the smooth conversation or produce one that goes awry. At this point, we can train a machine learning classifier to identify whether a conversation is going awry and, if so, for what reason.

Why not use ChatGPT to analyze a conversation instead of training a ML model? After all, we can fine-tune ChatGPT on our generated corpus of healthy vs. pathological conversations and use it for classification. The answer is that there are some cases where we would like an offline model that can be used with less computational resources. Moreover, ChatGPT still lags behind other models in terms of availability and computational resources.

It is important to emphasize that we are now in a better position to understand the context of conversations going awry. Imagine that you are an expert therapist listening to a married couple asking you for guidance. Your expertise is indispensable, but in real time and under the *cognitive load* of the conversation, you may not effectively monitor and regulate the session. Moreover, as a human being, you are biased, either by the theories on which you have been educated or unconsciously by hidden factors. For instance, you could be the therapist from "Good Will Hunting," and your inner representation of your abusive father could influence your perception of aggressive male figures. Bias is almost inevitable. Again, what happens if you are the child from Bateson's anecdote who became a psychotherapist? In this case, the wife might remind you of your mother, who forced you to eat spinach …

Imagine a context-based AI system called "Freud AI" as your virtual assistant. At a certain point, and when listening to an allegedly innocent utterance produced by the wife, your virtual assistant signals that the conversation is going awry. It further explains the reason: skipping or avoiding difficult topics. The husband has raised a difficult topic and expects his wife to address this difficulty. The wife responds by skipping her husband's utterance and shifting to another topic. Using contextual analysis, the utterances may gain meaning and signal that the therapist may want to intervene by pointing to the dangerous road the conversation is taking. In the context of the husband's expectation and the wife's response, ignoring this expectation may help you better understand why the husband responds angrily. Contextual understanding of conversations going awry may be illustrated using examples from cinema, TV, and theatre. I shall start with "Scenes from a Marriage," a masterpiece created by Ingmar Bergman in 1973.

7.2   Scenes from a Marriage

I decided to imitate a conversation from this TV series using the following prompt:

Prompt 7.4
Scenes from a Marriage is a 1973 Swedish television miniseries, written and directed by Ingmar Bergman.

##

Here are the scenes [HERE I INSERTED THE SCENES FROM WIKIPEDIA]:

##

Produce a conversation going awry between Johan and Marianne.

ChatGPT produced the following conversation:

Johan: "Marianne, I've been thinking... about us, our past, and where we stand now."
Marianne: (Pausing, looking up from her book) "That sounds serious. What about us?"
Johan: "I feel like we've lost something vital between us. Our connection isn't what it used to be."
Marianne: "Is this about Paula again? I thought we were past that."
Johan: "No, it's not just Paula. It's everything. It's like we're together but not really with each other."
Marianne: (Sighs) "Johan, I'm trying. I've been in therapy, working on myself for us. But it seems like you're always looking for something else, something beyond what we have."
Johan: "But don't you see? That's just it. I feel like you're more present when you're away from me. When we're together, it's as if we're just going through the motions."
Marianne: "So, what are you saying? That it's my fault? That I'm not trying hard enough?"
Johan: "No, that's not what I meant. I just... I don't know how to bridge this gap between us."
Marianne: (Frustrated) "Maybe there's just too much water under the bridge, Johan. Maybe we're trying to hold onto something that's long been lost."
Johan: "So, what? We just give up? After everything?"
Marianne: "I don't know, Johan. I really don't know."

Now, let me analyze the conversation to understand better *when*, *how*, and *why* the conversation is going awry. I focus on a single dimension of the context: expectations. I changed the names of Johan and Marianne to avoid any association with the original drama and used the following prompt:

Prompt 7.5
Expectations describe the anticipated response of a participant in a dialogue to a given utterance.

##

Guess Dana's expected response to the following utterance generated by Joe: "Dana, I've been thinking… about us, our past, and where we stand now."

##

Output in bullet points briefly describing the expected response regarding speech acts.

ChatGPT proposed the following speech acts:

Acknowledgment of Joe's thoughts: Dana acknowledges that Joe has been reflecting on their relationship, indicating that she is open to hearing more about his thoughts and feelings.

Expression of interest: She expresses interest in understanding what Joe has concluded from his reflections, showing willingness to engage in a deeper conversation about their relationship.

Sharing personal reflections: Dana might share her reflections on their past and current state of the relationship, contributing her perspective to the dialogue.

Expression of feelings: She communicates her feelings about their relationship, which could range from happiness and concerns to uncertainties, depending on her current emotional state and evaluation of their relationship.

Inquiry about the future: Dana may ask Joe about his thoughts on the future of their relationship, indicating a desire to understand his intentions or expectations.

Seeking clarity or resolution: If Joe's initial statement implies any ambiguity or concern, Dana might seek clarity or propose to discuss potential resolutions or next steps.

Marianne's actual response was (Pausing, looking up from her book), "That sounds serious. What about us?" Next, I used the following prompt:

Prompt 7.6
Dana responded by saying: "That sounds serious. What about us?"

##

To what extent is her response in line with the above expectations?

##

Output in the form of a Likert scale ranging from 1 (not at all) to 5 (very much)

ChatGPT provided a full explanation and decided to rate her response as "3," meaning that it was moderately in line with expectations. I repeated the process for the rest of the dialogue using the following prompt and this time with the original names:

Prompt 7.7
For each utterance in the following dialogue:

(1) Provide titles of the three most likely responses the speaker expects in response to his or her utterance. Describe the titles in terms of speech acts.
(2) Provide a Likert scale score indicating the extent to which the response of the other speaker to the previous utterance is in line with expectations.

    ##

    The dialogue is [INSERT DIALOGUE HERE]:

    ChatGPT provided a complete output, but I would like to focus on two cases where the alignment between the expected response and the actual response was low:

Johan: "I feel like we've lost something vital between us. Our connection isn't what it used to be."

Expected Responses:

1. Acknowledgment of concern
2. Expression of feelings
3. Inquiry for specifics or examples

   Marianne's actual response: "Is this about Paula again? I thought we were past that."
Likert scale: 2 (Somewhat in line, as Marianne does acknowledge the concern indirectly by referencing a past issue, but it does not align well with expressing feelings or inquiring for specifics related to Johan's general concern.)
   As you can see, Johan addresses a painful issue in his marriage. As a response, he probably expects at least some acknowledgment of concern. However, Marianne's actual response aligns with her concerns as she knows Johan had an affair with a younger woman called Paula. She forms a break in the conversation by guessing Johan's "true" intention and dismissing his present concerns as a past and closed issue. The next example involves the following utterance:

Marianne: "(Sighs) Johan, I'm trying. I've been in therapy, working on myself for us. But it seems like you're always looking for something else, something beyond what we have."

Expected Responses:

1. Recognition of effort
2. Reassurance or emotional support
3. Reflection on personal expectations or desires

Johan's actual response: "But don't you see? That's just it. I feel like you're more present when you're away from me. When we're together, it's as if we're just going through the motions."

Likert scale: 2 (Somewhat in line, as Johan does reflect on his feelings and the state of their relationship, but it lacks direct recognition of Marianne's efforts or offering reassurance.)

In this case, Marianne emphasizes her efforts to resolve their conflicts but also expresses frustration. She may expect recognition of her efforts, but Johan's response is disappointing as it lacks any recognition. An artificial therapist seeking to understand and indicate to the couple how the conversation is going awry can use our guidelines for contextual analysis. It can identify how an important aspect of context—prior expectationsmight lead the conversation to escalate. In such a dialogue, violated expectations will contribute significantly to the participants' sense of miscommunication and helplessness. The pathology expressed in the conversation is the pathology of missing an important aspect of the context: the expectations of the interlocutors. In the next section, I present another example from the famous play Pygmalion.

7.3 Context in "Pygmalion"

In George Bernard Shaw's "Pygmalion," a significant conversation goes awry in Act V, following the success of Eliza Doolittle at an ambassador's garden party, which proves that Professor Henry Higgins' experiment succeeded in transforming Eliza from a Cockney flower girl into a lady through lessons in phonetics and etiquette. The conversation occurs in Higgins' laboratory, where Eliza confronts Higgins about his treatment of her now that the experiment is over.

Feeling used and discarded, Eliza criticizes Higgins for not caring for her as a person but only as an experiment. Higgins is baffled by Eliza's ingratitude and inability to understand that he treats everyone with the same disregard, not just her. The conversation escalates as Eliza asserts her independence, claiming she no longer needs Higgins.

This conversation goes awry because each character has a fundamentally different expectation and understanding of the other. Higgins sees his experiment as a triumphant scientific achievement, while Eliza feels personally abused, and this leads to an emotional and philosophical clash. It showcases the transformation of Eliza, not just in terms of social class, but also in terms of self-awareness and independence, highlighting the play's exploration of social mobility, identity, and the nature of personal relationships. Here, I would like to focus on a single exchange. When addressing Colonel Pickering, Eliza says:

I know. I am not blaming him. It is his way, isn't it? But it made such a difference to me that you didn't do it. You see, really and truly, apart from the things anyone can pick up (the dressing and the proper way of speaking, and so on), the difference between a lady and a flower girl is not how she behaves, but how she's treated. I shall always be a flower girl to Professor Higgins because he always treats me as a flower girl and always will, but I know I can be a lady to you because you always treat me as a lady and always will.

Regardless of his allegedly indifferent approach, Higgins responds angrily, saying:

"She will relapse into the gutter in three weeks without me at her elbow."

What motivates this inappropriate, angry, almost vicious response is not entirely clear. It is true that Professor Higgins is a rude person who seems indifferent to human feelings. However, he treated Eliza as an object of experimentation, and one cannot accept gratitude from an object, whereas his response seems to express deep insult, as if he were expecting gratitude. So, his furious response is hard to explain without understanding Eliza's "speech" with its contextual dimensions. I used our contextual analysis and ChatGPT produced the following observations:

ELIZA's speech reflects on personal identity, social perception, and the impact of others' treatment on one's self-view. She contrasts the way Professor Higgins treats her, which perpetuates her identity as a flower girl, with the way another character treats her, elevating her to the status of a lady. This difference in treatment fundamentally alters her self-perception and societal role. Key themes in her utterance include:

- The importance of how others treat one in shaping self-identity.
- The distinction between social roles is based on treatment rather than inherent qualities or behaviors.
- A reflection on personal transformation and the role of external validation.

The themes identified by ChatGPT provide us with several hypotheses for explaining Higgins' furious response. Let me focus on the idea that social roles are determined by treatment rather than by inherent qualities. Professor Higgins is a clear example of a "gentleman" who does not have to work for a living. His social status supports his arrogance and self-assurance. As a member of the upper classes, he feels that he can behave as he likes. Here, Eliza challenges one of Higgins' most important principles by arguing that who you are, the basic foundation of your "self," is not determined by what you are but by how others treat you. Higgins' furious response raises the hypothesis that our expert professor is not as self-confident as one might think. Given these themes, let us identify which of the one hundred contextual statements align most closely with Eliza's experience and perspective as expressed in her utterance:

1. **The situation involves someone trying to convince someone of something**. While this may initially seem relevant because Eliza is explaining her feelings and trying to convey her understanding of her social standing, it is more of a reflection than an attempt to convince.
2. **It is a situation where someone is criticizing**. This is indirectly relevant because Eliza reflects on criticism and its effects, though her focus is more on treatment than on direct criticism.
3. **It is a situation where self-control is necessary**. Eliza's measured response, ability to articulate her feelings, and the distinctions in her treatment suggest self-control in navigating her social environment.
4. **It is a situation where someone needs or desires reassurance**. Eliza indirectly seeks reassurance about her identity and worth, as is further implied in her recognition of different treatments by different people.
5. **It is a situation that could make people tense and upset**. Eliza's reflection indicates underlying tension and upset feelings about the way Higgins treats her, as compared to the way others treat her.
6. **It is a situation where people are comparing themselves to each other**. Eliza compares the way individuals perceive and treat her, insofar as it direct affects her self-identity.
7. **It is a situation where someone is being pressured to conform to the actions of others**. This reflects Eliza's relationship with Higgins, who pressures her to conform to the societal expectations of a flower girl.
8. **It is a situation where someone needs to be taken care of**. Eliza's speech reflects a desire to be treated with care and respect, which she associates with being treated as a lady.
9. **It is a situation where members of a group respect and admire someone**. This is relevant to Eliza's perception that being treated as a lady by someone makes her feel respected and admired, in contrast to the way Higgins treats her.

Among these, the statement that best encapsulates the essence of Eliza's speech is:

It is a situation where people compare themselves to each other (#63).

This statement directly aligns with Eliza's reflection on how different treatments by different individuals lead her to compare herself with either a lady or a flower girl, significantly impacting her self-identity and perception. ChatGPT suggests that the context of Eliza's "speech" is social comparison. When asked to identify Eliza's expectations, ChatGPT provided us with the following:

1. **Acknowledgment**: Eliza expects an acknowledgment of her feelings and perspective. This acknowledgment would validate her experiences and the impact of Higgins' behavior on her self-perception. The speech act here is "expressive," where Higgins would acknowledge the truth in Eliza's observations and possibly express understanding or sympathy.

2. **Apology**: Given the context, Eliza might also expect an apology for the demeaning treatment she has endured. An apology would be another form of expressive speech act, signifying regret for actions taken and their impact on her.
3. **Change in behavior**: Beyond verbal acknowledgment or apology, Eliza likely hopes for a tangible change in the way Higgins treats her, reflecting a recognition of her growth and humanity. This would not be a speech act per se but a consequent action influenced by the conversation. If verbalized by Higgins, it could be a "declarative" speech act, where he commits to altering his behavior.

Eliza expects Higgins to acknowledge her status and see her, not as an object of experimentation, but as a human being. Treating her like a "flower girl" irrespective of the change she has undergone is a wrongdoing, an act requiring an apology and even a change in behavior. Here, we can see a conversation going awry because of different expectations. For the British gentleman, the idea of apologizing to a Cockney flower girl from the "gutter" is unthinkable. An apology is made for superiors or equals; a gentleman cannot be expected to apologize to a girl from a lower echelon of society. Here, two contexts converge, explaining Higgins' furious response and why the conversation is going awry. A flower girl from the "gutter" has been transformed into a lady. Her transformation is not superficial but deep and permanent. She will not "relapse" into the "gutter." If this is the case, she is a *real* lady, which means that asking the gentleman to apologize is legitimate, even from his arrogant perspective. Understanding the social comparison aspect of the context and the different expectations of the participants, we are now in a better position to understand Professor Higgins' fury. His experiment succeeded beyond his expectations. Straw turned into gold, and a flower girl from the gutter turned into a lady. Trusting his mastery of alchemy, Higgins should himself have believed that the transformation was not superficial. However, his behavior indicates that under his self-confident façade and sense of victory and achievement, there is a deep layer of anxiety, together with only a superficial understanding of humans beyond the mechanics of their maneuvers. The following example analyzes another conversation that is going awry.

7.4   Context in "American Beauty"

"American Beauty"[1] is a famous drama featuring Kevin Spacey as Lester Burnham, a middle-aged man who becomes disillusioned with his dull, suburban existence and develops an obsession with his daughter's teenage friend, leading to dramatic changes in his lifestyle and relationships. In one of the scenes, a family dinner slips into chaos. This scene is particularly significant because it encapsulates the film's

---

[1]https://en.wikipedia.org/wiki/American_Beauty_(1999_film).

themes of suburban despair, family communication breakdowns, and the characters' struggles with their identities and desires.

In the scene, Lester is at dinner with his wife Carolyn (played by Annette Bening) and their daughter Jane. The tense atmosphere highlights the family's strained relationships and lack of genuine communication. Lester, who is in the midst of a mid-life crisis, attempts to lighten the mood and connect with his family. However, his efforts are met with hostility, particularly from Carolyn, who is deeply unhappy with her life and marriage. The tension escalates as Lester rebels against Carolyn's control and the superficiality of their existence, leading to a confrontation that symbolizes Lester's growing rejection of his life. The scene is pivotal because it marks a turning point in Lester's transformation, as he begins to assert his desire for authenticity and happiness over the facade of a perfect suburban life. This dinner scene is often analyzed for its portrayal of the family's dysfunction and use of a mundane setting to explore deeper existential and emotional themes. The fight that erupts is not just about the immediate frustrations of the characters but also about their unfulfilled desires, their fears, and the societal pressures that shape their lives.

I asked ChatGPT to generate a conversation along the lines of the chaotic dinner conversation in "American Beauty," albeit with different names:

Juda: "Sarah, could you please pass the salad?"
Sarah: (Continuing to talk about her workday with enthusiasm) "…and then I said, that's not just a good deal, it's the best you're going to get!" (Laughs, ignoring Juda's request)
Linda: (Sits quietly, observing her parents, clearly uncomfortable with the growing tension)
Juda: "Sarah, the salad, please."
Sarah: (Pauses, looks at Juda with irritation) "You know, if you spoke more kindly, maybe you'd be happier with the responses you get."

I began by analyzing the first three lines without the explanations and Linda's perspective:

Juda: "Sarah, could you please pass the salad?"
Sarah: "…and then I said, that's not just a good deal, it's the best you're going to get!"
Juda: "Sarah, the salad, please."

The ambiance was identified as:

- DISTRACTED
- UNFOCUSED
- CASUAL

Now, I added Linda's response to the analysis and got:

- TENSE
- DEFENSIVE
- FRUSTRATED

With just one added line of conversation, the dinner's atmosphere/context has changed into tense, defensive, and frustrated. We now add the final part of the conversation without including the explanations in the brackets or Linda's part:

Juda: (Feeling ignored and frustrated, stands up abruptly, knocking his chair back slightly) "I just wanted the salad!" (He reaches across the table, grabbing the salad bowl forcefully)
Sarah: (Startled and angry) "There's no need to be dramatic! You're setting such a bad example for Linda."
Linda: (Wishes she could disappear, the air thick with tension, as her parents' argument escalates over something as trivial as salad, highlighting the deeper issues in their relationship)

And the atmosphere is now:

- TENSE
- CONFRONTATIONAL
- IRRITATED
- DEFENSIVE

The atmosphere has shifted from distracted, unfocused, and casual into tense, defensive, and frustrated, and, in the last part, it has become confrontational and irritating. There is nothing informative about this simple contextual analysis for those who saw the movie. The conversation deteriorates … However, even this simple contextual dimension of the situation, its ambiance, seems to provide a great deal of information about the changing context.

In real time, we are much better at *experiencing* than *reflecting*. Therefore, it is difficult for us to *monitor* and *regulate* a conversation in real time. An intelligent system can be trained to measure one contextual aspect of the situation: its ambiance. In a way that is quite intuitive to people and the manner in which they experience situations, the system can inform us that the *atmosphere* has changed.

Some participants' behaviors can be better understood and regulated in a changing context. Imagine a board meeting opening with niceties and congratulations while unnoticeably slipping into a poisoned atmosphere. An AI system observing the situation can inform the CEO that the context is changing and that he must do something to prevent the meeting from deteriorating further.

In this chapter, I have introduced the benefits of AI-supported context analysis for monitoring human conversations. I emphasized our shortcomings in processing social data in real time and pointed to the consequences such shortcomings might have for monitoring and regulating an interaction. Whether we are therapists seeking to understand the changing contexts of marital therapy or managers seeking to understand board meetings, AI-supported context analysis may significantly leverage our capabilities. As communication breakdown is so common, understanding the context of conversations is a challenge with clear implications.

References

1. Becker, A.L.: Text-building, epistemology, and aesthetics in Javanese shadow theatre. Dispositio **5**(13/14), 137–168 (1980)
2. Zhou, P., et al.: Commonsense-focused dialogues for response generation: an empirical study. arXiv preprint arXiv:2109.06427 (2021)

Chapter 8
The Context of Social Norms

Abstract As previously explained, one cannot understand social norms and their violations without considering the context. Here, I further explain this idea and show how we can delve deeper into social norm violations using automatic context analysis. More specifically, I focus on a scene from the comedy "Borat" (https://en. wikipedia.org/wiki/borat), where the provocative Sacha Baron Cohen ruins a pleasant dinner party in the American South.

8.1 Social Norms Are Contextual

What if I told you I saw Moshe walking naked on the beach? Is this surprising news? Does this behavior indicate the violation of a social norm? It all depends on context. If Moshe is known as a nudist who regularly attends a nudist beach, it is not surprising to hear that he walked naked on the beach, and this behavior does not involve a violation of a social norm. However, if Moshe appeared nude on a *public* beach, he violated a social norm.

What if I tell you that Zhang Wei saw Li Hong in the office and said, "Hi Hong"? Is this utterance an example of a norm violation? In Chinese culture, it is customary to use titles followed by the surname when addressing someone professionally, especially if they are your superior. This could include titles such as "经理" (Jīnglǐ, Manager), "总监" (Zǒngjiān, Director), or simply "先生" (Xiānsheng, Mr.) or "女士" (Nǚshì, Ms.) followed by their surname. The direct use of a first name without a title can be seen as overly familiar or disrespectful, particularly in a formal or professional context. If Li Hong is Zhang Wei's boss, Wei's impolite greeting is a norm violation.

These two examples show that social norms and their violations are deeply contextual. An isolated piece of verbal or non-verbal behavior is meaningless unless context is considered. Social norms are the unwritten rules that govern human behavior within a society or group. They involve the shared expectations that guide us by providing patterns of behavior and communication that support social coordination and stabilize social structures. Social norm violations occur when an individual acts in a way that does not follow these unwritten rules. And these rules

Y. Neuman, *AI for Understanding Context*, SpringerBriefs in Computer Science, https://doi.org/10.1007/978-3-031-64210-4_8

are contextual. In a profound sense, these violations are similar to the pathology of language discussed by Becker. In both kinds of situation, a given piece of behavior is *incoherent* with or *contradictory* to some constraints (i.e., the context). In the context of a nudist beach, walking naked is coherent behavior. However, such behavior is pathological in the context of a public beach. In the context of friends having a party, an informal way of approaching your friends is coherent with the context. However, greeting your boss on your first day at the office by saying: "Hi Joe, what's up bro?" is incoherent with the context.

The use of ChatGPT to contextualize behaviors may help us to understand social norms and their violations. Let me illustrate this point using the film "Borat." This is a hilarious comedy featuring Sacha Baron Cohen as Borat Sagdiyev, a fictional Kazakh journalist traveling to the States to make a documentary film on American society. Unfortunately, Baron Cohen chose Kazakhstan as a country that epitomizes a primitive and barbaric society. This is, of course, not the case. Regardless of this disrespectful attitude, the comedy makes us laugh. Moreover, through outrageous and often provocative interactions with unsuspecting Americans, the film epitomizes themes of cultural misunderstanding of social norms and their violations. To identify norm-violating situations, I gave ChatGPT the movie plot and the following instruction:

Prompt 8.1
Identify the top five social norm violations in "Borat."

##

Output in bullet points describing (1) the exact situation expressing norm violation and (2) the violated norm.

ChatGPT identified the following situations:

1. **Situation**: Borat wears a bright green mankini on a public beach.

   Violated Norm: Dress code and decency in public spaces.

2. **Situation**: Borat brings a bag of his feces to the dinner table at a Southern dinner party.

   Violated Norm: Basic hygiene and etiquette at the dining table.

3. **Situation**: Attempting to kidnap Pamela Anderson to make her his wife.

   Violated Norm: Consent, personal autonomy, and legal norms against kidnapping.

4. **Situation**: Borat's anti-Semitic remarks and actions, including the "Running of the Jew" sequence.

   Violated Norm: Respect for all ethnicities and religions; promoting hate speech.

5. **Situation**: Borat and his producer engage in a naked fight that spills into a mortgage brokers' conference.

   Violated Norm: Public decency, personal space, and professional conduct.

Let me start with the second disgusting situation where Borat brings a bag of his feces to the table at a Southern dinner party. This behavior seems to result from his cultural background and the lack of modern toilets in his home village. One might expect it to violate any civilized norm regardless of context. However, Borat's behavior suggests that, in his culture, such behavior is socially legitimate. Can we imagine contexts where bringing feces in a bag would be normative? Imagine a medical setting where feces are analyzed for health or research purposes. This is a context where the collection and handling of fecal matter would be considered quite normal. However, this behavior is totally *out of context* at a *dinner party*. The set of circumstances described as a *dinner party* entails situational and contextual dimensions. This set of circumstances provides a unique setting, whether a social event, workplace, or private space, which influences the behaviors and interactions of the people involved. Here is a possible prompt for identifying the situation:

Prompt 8.2

Objective: Your task is to distill the essence of a given scenario or situation into a concise title that encapsulates the broad or general context. The focus should be on identifying and summarizing the primary setting or activity in its most encompassing terms rather than highlighting specific actions, details, or participants.

Instructions

1. **Read and analyze the scenario/situation**: Carefully understand all elements presented, including the setting, the participants, and notable events or actions.
2. **Identify the general context**: Determine the scenario's broadest or most encompassing aspect. This should reflect the primary setting or activity, focusing on the overall environment or event rather than the intricacies of what takes place within it.
3. **Craft a concise title reflecting the general situation**: Your title should succinctly summarize the overarching context or activity that defines the scenario. The goal is to capture the essence of the situation in broad terms, avoiding specific details or elements that are not essential to understanding the general nature or setting of the scenario.

Guidance

- **Aim for broad categorization over specificity**: If the scenario involves unique or specific actions within a common setting, your title should prioritize the setting. The title should be broad enough to encompass any specific activities without detailing them.
- **Focus on the essential aspect**: Consider what aspect of the scenario is vital for grasping its general nature or setting. The title should make this aspect immediately apparent to the reader.

Examples for Clarity Given scenario/situation: "During a corporate team-building retreat, employees participate in various outdoor activities, including a trust fall exercise, where they take turns falling backward into the arms of their colleagues to build trust. The event occurs in a large park with facilitators guiding the activities."

- **Desired Output**: "Corporate team-building retreat"
- **Given scenario/situation**: "Borat brings a bag of his feces to the dinner table at a Southern dinner party."
- **Desired Output**: "Dinner party."

Using this prompt, I analyzed the following behavior: Borat wears a bright green mankini on a public beach. ChatGPT produced the following title: "Public beach outing." In the "public beach outing context," walking with a mankini is an anomaly …

The general context identified by the above prompt entails a set of expectations regarding the situational and contextual dimensions previously discussed. Let me identify the contextual dimensions of the second situation to explain how we may use ChatGPT to process social norms.

The situation is: "Borat brings a bag of his feces to the dinner table at a Southern dinner party." The general situation identified by ChatGPT is "Dinner party," and we already know that Borat's behavior violates norms of "Basic hygiene and etiquette at the dining table." I asked ChatGPT to analyze the general situation of a dinner party to expose its situational and contextual dimensions:

Prompt 8.3
Extract from the following SITUATION the dimensions <WHO> <WHY> <WHEN> <WHERE> <OBJECT or WHAT> <HOW>.

##

Provide your best guess if a <SITUATIONAL DIMENSION> is not explicitly mentioned.

##

The situation is: a dinner party.

##

<WHO>: The individuals in the situation. <WHY>: The underlying motivation or intention underlying the participants' behavior. <WHERE>: The physical location of the situation. <WHEN>: The specific time or period mentioned or implied by the situation. <OBJECTS or WHAT : The tangible or intangible objects present or relevant in the situation. <HOW>: Information about actions and procedures in the situation. <SITUATIONAL DIMENSION>: <WHO>, <WHY>, <WHEN>, <WHERE>, <OBJECTS or WHAT>, <HOW>.

##

OUTPUT: A list of parameters for each <SITUATIONAL DIMENSION>.

ChatGPT produced the following output:

For the general situation of "a dinner party," let us extrapolate the dimensions based on common understanding and typical scenarios associated with such an event:

<WHO>: Hosts (individual or couple organizing the dinner), guests (friends, family, colleagues), and possibly service staff (if it is a formal dinner party).
<WHY>: To socialize, celebrate a special occasion (birthday, promotion, holiday), or enjoy good food and company.
<WHERE>: Typically at the host's home, but could also be at a rented venue or in a restaurant's private dining room.
<WHEN>: Evening is most common for dinner parties, often on weekends or special occasions. The specific date might depend on the reason for the gathering.
<OBJECTS or WHAT>: Dinner table, chairs, dinnerware (plates, cutlery, glasses), food and beverages, possibly decorations if it is a celebratory or themed dinner party.
<HOW>: Guests arrive, are greeted by the host, and may be offered a drink. They engage in conversation, take their seats at the dinner table, and are served multiple courses, or help themselves in a more casual buffet style. The evening typically includes dining and socializing, with the host facilitating introductions and conversations to maintain a pleasant atmosphere.

What is the expected *ambiance* at a dinner party? I used the following prompt:

Prompt 8.4
Ambiance is the mood, character, quality, tone, atmosphere, etc., particularly of an environment or milieu.

##

Describe the ambiance expressed in the following situation: a dinner party.

##

Output is in the form of bullet points with one word for each bullet point (e.g., SERENE, WARM).

ChatGPT explained that the ambiance of a dinner party is:

- WELCOMING
- FESTIVE
- INTIMATE
- SOCIABLE
- ELEGANT
- COZY
- LIVELY

I would like to add to our analysis the situational dimension of SCRIPT. The script typically involves a sequence of expected actions, behaviors, and speech acts in a given situation. It is an extended aspect of the HOW dimension that I described previously when defining the dimensions of a situation. I used the following prompt to identify the script of a dinner party:

Prompt 8.5
[Identify the script of a situation]

(A script is the sequence of actions, behaviors, and speech acts individuals are expected to follow in a given situation.)

Extract the script expressed in the following situation: a dinner party.

(If a script or an action is not explicitly mentioned, please provide your best guess based on the context).

##

Identify the individual responsible for each action in the sequence.

ChatGPT identified the following script of a dinner party:

Preparation (Host)
Planning the menu.
Inviting guests, often via phone, email, or physical invitations.
Cleaning and decorating the home or venue.
Preparing or cooking the meal, setting the table.

Arrival (Guests)
Arriving at the host's home or the venue, usually bringing a small gift (e.g., wine, flowers).
Greeting the host and other guests.

Welcome (Host)
Greeting guests as they arrive.
Taking coats or personal items.
Offering guests a drink or directing them to a seating area.

Socializing (Both)
Engaging in conversation, introductions if necessary.
Participating in any pre-dinner activities or entertainment provided by the host.

Seating for Dinner (Host & Guests)
The host signals when dinner is ready.
Guests may be directed to specific seats if there's a seating plan.

Dining (Both)
The host serves the food (or guests serve themselves in a buffet setup).
Enjoying the meal while engaging in conversation.
Complimenting the host on the meal.

Post-Dinner Activities (Both)
The host may serve coffee, tea, or dessert.
Engaging in post-dinner entertainment, games, or continued conversation.

Departure (Guests)
Guests indicate readiness to leave.
Thanking the host for the evening.
The host sees guests out, thanking them for coming.

Follow-Up (Guests & Host)
Guests may send a thank-you note or message the following day.
The host may reach out to express enjoyment of the evening.

This script outlines the general expectations and sequence of actions typical for a dinner party, highlighting the social norms and behaviors anticipated by hosts and guests. At this point, I should have general situational knowledge about a dinner party. To this knowledge, I added the contextual dimensions using the following prompt:

Prompt 8.5
Shared knowledge describes the information, understanding, or experiences expected from a situation. Shared knowledge is necessary for understanding the situation.

##

Describe the shared knowledge required to understand the following situation: a dinner party.

##

Describe the shared knowledge in terms of information, understanding, and experiences.

The output I got was as follows:

Information

1. **Etiquette rules**: Knowledge of basic social etiquette, such as punctuality, dressing appropriately, and table manners.
2. **Host–guest dynamics**: Awareness of the roles and expectations of hosts and guests, including the host's responsibility to make guests feel welcome and the guests' responsibility to be courteous and appreciative.
3. **Cultural or thematic expectations**: Knowledge of relevant customs or thematic elements is expected if the dinner party has a specific theme or is influenced by cultural traditions.

Understanding

1. **Social cues and non-verbal communication**: Recognizing and responding appropriately to social cues, such as when to start eating, how to signal you are

finished with your meal, and understanding body language and facial expressions to gauge the comfort and engagement levels of others.

2. **Conversation etiquette**: Understanding how to engage in polite and inclusive conversation, avoiding controversial or inappropriate topics while showing interest in fellow guests.
3. **The purpose of the gathering**: Grasping whether the dinner party is for a special occasion, a casual get-together, or has a particular aim (e.g., celebrating an achievement or a farewell dinner) to adjust one's expectations and behavior accordingly.

Experiences

1. **Participation in previous social gatherings**: Previous experiences attending or hosting social gatherings help individuals navigate the dynamics of a dinner party, from arrival to departure.
2. **Culinary appreciation**: An appreciation for food and drink, understanding the effort behind meal preparation, and the ability to engage in conversations about the cuisine being served.
3. **Adaptability to social norms**: Experiences that have taught adaptability and sensitivity to different social norms and customs, especially in diverse cultural settings.

Next, I identified the coherence dimension using the following prompt:

Prompt 8.6

In understanding a situation, we can differentiate between inferences and hypotheses.

1. Inferences involve drawing logical conclusions from or interpreting information explicitly stated or implied within the situation.
2. In understanding a situation, a hypothesis can be seen as a tentative explanation or interpretation of certain elements or events within the situation. It involves proposing a possible relationship, explanation, or outcome based on the evidence or information presented in the situation.

##

Produce inferences and hypotheses for the following situation: a dinner party.

##

Inferences and hypotheses should focus only on the psychological and social aspects of the situation.

##

ChatGPT suggested that for the situation of a dinner party, we should explore both inferences and hypotheses related to its psychological and social aspects:

Inferences

1. **Social bond strengthening**: Sharing a meal in a social setting, like a dinner party, is implicitly understood to strengthen participant bonds.
2. **Desire for socialization**: The organization and attendance of a dinner party imply a desire for social interaction, indicating that participants value social connections.
3. **Cultural and social norms**: The setting and conduct of the dinner party reflect adherence to particular cultural and social norms, suggesting a mutual understanding of expected behaviors.

Hypotheses

1. **Effect of ambiance on social interactions**: The ambiance of the dinner party, such as lighting, music, and table-setting, may significantly influence the mood and conversational dynamics among guests. A warm and inviting atmosphere is hypothesized to facilitate more open and engaging interactions.
2. **Impact of diverse guest composition on social cohesion**: Introducing guests from diverse backgrounds or differing viewpoints may enhance the event's social richness, leading to more stimulating conversations, or pose challenges to social cohesion if not managed sensitively.
3. **Role of food in social connection**: The quality, variety, and presentation of food can serve as a conversation starter and bond-forming element, hypothesizing that meals prepared with care and thoughtfulness contribute to a more positive and memorable social experience.
4. **Influence of host's behavior on guest comfort**: The host's ability to manage conversations, introduce guests to one another, and ensure everyone feels included is hypothesized to directly impact the overall comfort and satisfaction of the guests, potentially influencing the success of the dinner party.

The expectations were identified through the following prompt:

Prompt 8.7
Expectation describes the anticipation or belief that something will happen in the FUTURE.

##

Describe the participants' expectations in the following situation: a dinner party.

And ChatGPT suggested that:

In a dinner party, the participants' expectations can vary depending on their roles (e.g., host vs. guest) and the nature of the dinner party (e.g., formal vs. casual). Here are some common expectations:

Host Expectations
Guest attendance: The host expects the guests to arrive on time and attend as confirmed.

Appreciation and enjoyment: The host anticipates that guests will appreciate the effort put into organizing the party, including the food, ambiance, and overall experience, and that they will enjoy themselves.

Social engagement: The host hopes that guests will converse with each other, contributing to a lively and pleasant atmosphere.

Respect for property: The host expects guests to treat the home or venue and its belongings respectfully, adhering to any house rules.

Guest Expectations

Welcoming environment: Guests expect to be welcomed warmly by the host and feel comfortable throughout the event.

Good food and beverages: Guests anticipate being served delicious food and drinks appropriate to the occasion and the time of day.

Social interaction: Guests look forward to engaging with other attendees, expecting interesting and respectful conversations. They may also expect to meet new people or catch up with old friends.

Clear communication: Guests expect clear communication from the host regarding the details of the party, such as time, location, dress code, and whether they should bring anything.

Shared Expectations

Enjoyment and relaxation: Both hosts and guests expect the dinner party to be an enjoyable and relaxing occasion, allowing everyone to take a break from their daily routines.

Etiquette and manners: There is a mutual expectation that all participants will adhere to social norms and manners appropriate to the setting, such as polite conversation, offering thanks to the host, and helping with minor tasks if needed.

Safety and comfort: In light of health concerns or dietary restrictions, participants expect that the environment will be safe and that any such restrictions will be considered in the planned food and activities.

Given the contextual information gathered so far, I asked ChatGPT to identify the contextual dimensions *violated* by Borat's behavior. ChatGPT detailed these violations as follows:

Violated Situational Dimensions

<WHO>: This behavior disrupts the host and guests' roles by introducing a socially unacceptable element.

<WHY>: The underlying motivation of a dinner party is to socialize and enjoy each other's company in a pleasant environment, which is directly contradicted by Borat's action.

<WHERE>: A dinner table, especially in a Southern dinner party context, is a place for eating and socializing, not for the introduction of inappropriate materials.

<OBJECTS or WHAT>: Items brought to the table typically include food, drinks, and possibly decorative items, not objects that are universally considered unsanitary.

<HOW>: The actions and procedures expected at a dinner party involve dining, conversing, and perhaps some entertainment, all of which are disrupted by this act.

Violated Ambiance

WELCOMING, FESTIVE, INTIMATE, SOCIABLE, ELEGANT, COZY, LIVELY: Each of these qualities of ambiance is negatively impacted, as the act introduces elements of shock, disgust, and discomfort, completely altering the mood from positive to negative.

Violated Script

Each step of the typical dinner party script is undermined, from the welcome to socializing, dining, and departure. Borat's behavior introduces an unexpected and unwelcome action that falls outside conventional or acceptable behavior at such gatherings.

Violated Shared Knowledge

Etiquette rules, host–guest dynamics, cultural and social norms: The shared understanding that dinner parties are occasions for pleasant social interaction is violated, as is the expectation for guests to contribute positively to the ambiance.
Social cues and non-verbal communication: Borat's action ignores social cues and the non-verbal expressions of discomfort it would undoubtedly cause.
The purpose of the gathering: This act detracts from any celebratory or social purpose of the dinner party.

Violated Expectations

Enjoyment and relaxation, etiquette and manners, safety and comfort: All these anticipations are disrupted. Guests and hosts alike expect a safe, comfortable environment conducive to enjoyment and relaxation, predicated on mutual respect and adherence to social etiquette.

Violated Inferences

Social bond strengthening, desire for socialization: The inference that a dinner party strengthens social bonds and fulfills a desire for socialization is contradicted. Instead, this act would likely result in alienation and discomfort.

Borat's action of bringing a bag of his feces to a Southern dinner party table starkly violates the foundational principles of social interaction, etiquette, and respect that define the situation of a dinner party, leading to a breakdown in the social fabric that should normally be reinforced by such gatherings. His act challenges social norms and taboos involving basic codes of hygiene and etiquette. The humor associated with Borat's provocative behavior comes from the fact that our little inner rebel takes pleasure in observing the violation of social norms without having to pay the price of *shame* or *embarrassment*. Using our contextual analysis, we may systematically and scientifically identify pathological behaviors resulting in shame or embarrassment. For example, I used the following prompt to identify the ambiance produced by Borat's shameful behavior:

Prompt 8.8

Ambiance is the mood, character, quality, tone, atmosphere, etc., particularly of an environment or milieu.
##

Describe the ambiance expressed in the following situation: Borat brings a bag of his own feces to the dinner table at a Southern dinner party.

##

Output is in the form of bullet points with one word for each bullet point (e.g., SERENE, WARM).

ChatGPT proposed that the ambiance is:

- SHOCKING
- UNCOMFORTABLE
- DISRUPTIVE
- BIZARRE
- INCONGRUOUS

The situation is shocking, uncomfortable, and disruptive for a good reason. It evokes disgust among the hosts and guests, far outside their expectations for normal dinner party behavior. The guests and hosts will likely experience shock and discomfort, as bringing feces to a dinner table violates universally recognized standards of cleanliness and respect in social settings.

8.2   Context and Social Emotions

Borat's behavior in the above-mentioned situation lacks any signs of guilt or regret. He does not seem to feel shame or embarrassment as his cultural background leads him to a different understanding of the situation and the context. Therefore, Borat cannot be blamed for his inappropriate behavior because he does not have bad *intentions*.

Surprisingly, those who have studied context, from Voloshinov to Becker, ignored or underestimated the emotional aspect of context. It is as if human emotions had nothing to do with the intellectual realm in which these scholars worked. This is unfortunate. Like other social mammals, humans are emotional creatures, and social emotions play a significant role in our lives.

To understand whether a certain piece of behavior expresses the pathology of misunderstanding context, we can use ChatGPT in two different ways. First, we can use a theory of social emotions. Sznycer et al. [1] suggest that social emotions are adaptive and functional in calibrating our behavior to self and others. For example, Sznycer et al. [2] argues that shame serves the adaptive function of defending against the social devaluation that results when negative information reaches others. Imagine someone lying to their best friend about a serious matter, which later causes significant harm or distress to the friend. When the truth comes out, the person who lied may experience shame for their dishonesty and the damage it caused to the trust and bond in their friendship. This shame stems from recognizing the violation of social norms about honesty and integrity in relationships. Knowing about this behavior, others may devaluate someone who cannot be trusted. Such behavior entailing distrust from others might have detrimental consequences for the person who violated the social norm. At this point, we should better understand the difference between shame and embarrassment and guilt and regret. ChatGPT can help us to understand the difference.

Shame

Internal and global: Shame is a deeper, more painful emotion, related to a person's feelings about themselves. It involves a sense of worthlessness or being funda-mentally flawed in some way. When someone feels shame, they are likely to perceive their entire self as bad, inadequate, or unworthy.

Focus on the self: Shame focuses on the self rather than the specific behavior. It implies a negative evaluation of one's core identity rather than just one's actions.

Long-lasting impact: Shame can have a long-lasting impact on an individual's self-esteem and self-image. It is more likely to be internalized and can affect one's mental health and relationships over time.

Embarrassment

External and specific: Embarrassment is typically a milder and more fleeting emotion than shame. It arises from specific actions or situations that are *socially awkward* or *inappropriate* but not necessarily indicative of one's character.

Focus on the behavior: Embarrassment is more about how others perceive one's actions in a specific context. It involves a sense of awkwardness or discomfort stemming from a social *faux pas* or mistake.

Short-term and situational: Embarrassment is usually short-lived and directly tied to a particular situation or behavior. The feeling generally subsides once the situation is resolved or enough time has passed.

In sum, shame is a more profound, encompassing feeling that pertains to one's sense of self and worth, often resulting from moral or ethical failings. In contrast, embarrassment is a temporary, situational response to a social mishap, focusing more on specific actions than intrinsic worth.

What is the difference between guilt and regret? These are emotions related to recognizing one's mistakes or wrongdoings, but they differ in focus, cause, and how they influence behavior. As explained by ChatGPT:

Guilt

Focus on action: Guilt arises when a person believes they have done something wrong or harmed someone else, *violating moral, ethical, or social standards*. It is closely tied to the act and the person's role in causing harm.

Moral or ethical dimension: Guilt has a strong moral or ethical component, involving the realization that one's actions have negatively impacted someone else's well-being or have breached one's moral code.

Motivation for amends: Guilt often motivates individuals to make amends or take *corrective action*. It involves a sense of responsibility for one's actions and a desire to right the wrong, possibly leading to apologies or efforts to repair the damage.

Regret

Focus on outcome: Regret is felt when a person wishes they had made a different decision, not necessarily because it was morally wrong, but because it led to an *unwanted outcome. It's more about poor outcomes than ethical violations.*

Broader than ethical concerns: Regret can arise from decisions that have nothing to do with ethics or morality, such as missed opportunities or poor judgment calls that lead to personal loss or inconvenience.

Reflective and forward-looking: While regret can lead to learning from one's mistakes, it often involves a reflective assessment of past decisions and how things could have been different. It's as much about what wasn't done as what was done.

Given these social emotions, a straightforward approach for identifying social norm violations in a given context may be as follows:

Prompt 8.9

Here is a situation: dinner party.

Here is a piece of behavior in a situation: Borat brings a bag of his feces to the dinner table at a Southern dinner party.

##

Identify which of the following emotions should be experienced by the agent performing the behavior and the participants observing the behavior.

##

The emotions are: anger, gratitude, shame, anger, embarrassment.

ChatGPT answered as follows:
For the behavior described in the situation where Borat brings a bag of his feces to the dinner table at a Southern dinner party:

Emotions Experienced by the Agent (Borat) Performing the Behavior

Shame: Typically, someone who performs a socially unacceptable act might feel shame; however, given Borat's character and intentions (to provoke and satirize), he may not experience this emotion as expected in real-life scenarios.

Embarrassment: Similar to shame, while a real person might feel embarrassment for such a socially inappropriate action, Borat's character is portrayed as largely immune to this due to his comedic and provocative nature.

Emotions Experienced by the Participants Observing the Behavior

Anger: Observers/participants at the dinner party may feel anger due to the disruption and disrespect shown towards the social norms of the setting.

Embarrassment: Guests are likely to feel embarrassment, not only on behalf of Borat for committing such a social *faux pas*, but also for themselves being part of a situation where social norms are being violated.

Shame: Some guests might feel a vicarious sense of shame, reflecting on how the behavior tarnishes the collective social experience of the dinner party.

We can see that ChatGPT "understands" the emotional aspects of the situation and can easily identify the emotions associated with social norm violations.

There is a second approach for identifying social emotions expected from a context. This approach draws directly on the context and the ambiance. I used the ambiance describing the general situation (i.e., dinner party) and the ambiance characterizing Borat's specific behavior (situation 2):

Prompt 8.10

Here are five words describing the ambiance of two situations:

Situation 1: WELCOMING FESTIVE INTIMATE SOCIABLE ELEGANT
Situation 2: SHOCKING UNCOMFORTABLE DISRUPTIVE BIZARRE INCONGRUOUS

##

Guess which social emotions best describe the situation based on the words describing the ambiance in each situation.

ChatGPT responded as follows:

Situation 1 Emotions

Happiness: The ambiance suggests a positive and joyful setting where participants feel happy and content.

Comfort: The welcoming and elegant environment implies a sense of comfort and ease among participants.

Connection: The intimate and sociable ambiance suggests a strong connection and feeling of belonging among individuals.

Situation 2 Emotions

Embarrassment: The shocking and uncomfortable nature points towards feelings of embarrassment among participants.

Anxiety: The disruptive and bizarre ambiance can induce anxiety or stress as participants navigate unexpected or uncomfortable situations.

Confusion: The incongruous nature of the situation will likely lead to confusion as participants struggle to make sense of the behavior or events unfolding.

We can see that automatically identifying the ambiance in the situation type, the particular behavior, and their associated emotions can give us another approach to identifying social norm violations.

8.3 Summary and Conclusions

Once, I observed a meeting[1] between the American Secretary of State Anthony Blinken and the Chinese Foreign Minister Wang Yi. Blinken opened the meeting with greetings and gratitude but almost immediately moved to a different issue:

"We'll also discuss our deep concerns with actions by China, including in *Xinjiang …*"

This mention of Xinjiang would mean nothing to most people, so understanding what happened in this meeting would be impossible for them. However, the "*deprivation of liberty of Uyghurs and other predominantly Muslim communities*" is of concern to some American progressive liberals, as indicated by the New Yorker article from which this quote was taken.[2] When publicly expressing "deep concerns" with Chinese actions in Xinjiang, Blinken violated basic social norms. First, this area is a Chinese territory with significant economic importance.[3] It is "China's largest natural-gas-producing region" (see footnote). A superpower, such as China, does not like others interfering with its internal affairs, specifically as it concerns a region with significant national interests. To understand this point, imagine the Chinese foreign minister saying:

We are deeply concerned about human rights in the States. "The United States leads the world in total number of people incarcerated, with more than 2 million prisoners nationwide (per data released in October 2021 by World Prison Brief)."[4] As a peaceful nation striving for social harmony, we are very concerned by these numbers. Moreover, as a nation with a socially oriented agenda, we are concerned by inequality in the States. *The World Economics Inequality Index* shows that the United States is ranked in the same category of inequality as countries like Bolivia and Mexico.[5] We are concerned by this inequality and would like to collaborate with our American counterparts to help them resolve this issue.

This imaginary statement has a clearly patronizing flavor, and one can easily imagine the American response. I doubt whether Blinken deliberately made this insulting public move. If this is the case, it is unfortunate because, apart from saluting his liberal base "concerned"[6] with human rights, nothing productive was initiated by or resulted from this violation of social norms. To the best of my knowledge, no improvement in human rights was subsequently observed, and nor

---

[1]https://asia.nikkei.com/Politics/International-relations/US-China-tensions/How-it-happened-Transcript-of-the-US-China-opening-remarks-in-Alaska.

[2]https://www.newyorker.com/news/q-and-a/why-hasnt-the-un-accused-china-of-genocide-in-xinjiang.

[3]https://en.wikipedia.org/wiki/Xinjiang#: ~ :text=In%201955%2C%20Xinjiang%20was%20administratively,natural%2Dgas%2Dproducing%20region.

[4]https://worldpopulationreview.com/country-rankings/incarceration-rates-by-country.

[5]https://www.worldeconomics.com/Rankings/Economies-By-Inequality.aspx.

[6]Although not concerned by the rights of Israeli Jews. See the hilarious Welcome to Columbia Untisemity: https://www.youtube.com/watch?v=rbfccVBo9tE.

was there any improvement in the relationships between the two superpowers. One may wonder whether, beyond the "experts" who advised Blinken, we could have developed a more "objective" intelligent system exposing the context of Blinken's speech and the dynamics of the meeting. Such a context-sensitive AI system could have informed Blinken that by priming the context of Xinjiang, he might push the meeting into the quicksand of confrontation.

References

1. Sznycer, D., Sell, A., Lieberman, D.: Forms and functions of the social emotions. Curr. Dir. Psychol. Sci. **30**(4), 292–299 (2021)
2. Sznycer, D., Tooby, J., Cosmides, L., Porat, R., Shalvi, S., Halperin, E.: Shame closely tracks the threat of devaluation by others, even across cultures. Proc. Natl. Acad. Sci. **113**(10), 2625–2630 (2016)

Chapter 9
Final Thoughts on AI for Understanding Context

Abstract This short concluding chapter explains why the idea of AI is presented in this book in the sense of Augmented Intelligence, why the book should be read as a first invitation to AI-based contextual analysis, and that in a highly dynamic and innovative field such as LLMs, emerging technologies can change our computational contextual analysis.

I was first introduced to the importance of context when I was a graduate student. An anthropologist, Zvi Bekerman, advised me to read Voloshinov, Bateson, and Becker, who advocate understanding human behavior by considering "context." These scholars have changed the way I understand meaning and interpretation. However, those were days when we lacked the technologies to support the complexity of a deep contextual analysis. By default, context-sensitive analysis remained within the boundaries of subjective interpretation with all the difficulties accompanying this approach, including the pitfall of "over-interpretation" and ideologically based interpretations.

For years, I have experimented with NLP technologies to model situations (e.g., [1]). However, the results were limited because the technology could not support the challenge of a deep contextual analysis. This all changed when Large Language Models (LLMs), specifically GPT, emerged, revolutionizing the field of AI and NLP. Here was an indispensable opportunity to merge old and sophisticated theories with new tools. At last, ideas that took into account the complexity of human behavior could be used in a way never seen before. This potential has been studied in some of my publications (e.g., [2]). However, my three years in a DARPA-sponsored project (2022–present) have given the most significant push to my research into context and AI.

The Defense Advanced Research Projects Agency (DARPA) supported a project titled Computational Cultural Understanding (CCU). In this project, I developed algorithms for identifying social norms and their violations (e.g., [3]) in various languages, from Chinese Mandarin to Russian. However, the project provided the participants with de-contextualized fragments for the analysis, and it is hard to develop intelligent systems for identifying social norms and their violations in a

Y. Neuman, *AI for Understanding Context*, SpringerBriefs in Computer Science, https://doi.org/10.1007/978-3-031-64210-4_9

conversation unless context is taken into account. On the other hand, no protocols were available to guide me in the search for a solution. This challenge brought me back to my long-lasting interest in contextual understanding. The present book emerged from painful reflections on this challenge.

In this book, I tried to explain the importance of context, to introduce ideas and theories proposing ways to model context, and most importantly, to show how we may use AI to leverage contextual understanding and support challenges that, in the past, remained the ultimate expression of human expertise. It is important to emphasize that the design of an intelligent system with a module for contextual understanding may significantly benefit from "basic" science and the theoretical insights of great minds, even if they lived a hundred years ago. This is why I explained the importance of context and used the insights of scholars like Voloshinov and Becker.

While promoting the thesis that AI may be helpful in understanding context, the book rejects any naïve perspective seeking to replace the human expert with the machine. The concepts and technologies presented in this book are not intended to replace humans but to empower them, similar to the way in which the invention of the piano empowered musicians from Bach to Faure.

The public discourse involving current AI is accompanied by fears. Some of these fears are fully justified, while others are the projections of our ignorance. However, the power of technology should not be dismissed because of fears. Imagine a musical talent with the stature of Bach, living before the first piano was invented. Sitting at home and being informed about this new technology, the piano, such a person might have replied by saying, "For God's sake! Why use a tool for creating music when we can sing so nicely? This satanic tool might replace humans!" Luckily, great musicians realized that the piano provides new music-making opportunities. The piano did not replace the musicians but empowered music. The present book should be read as one possible way of extending the tool kit available to researchers interested in how context can be automatically identified and used to meet various research and development challenges. Therefore, whenever I discussed AI, I used it mainly in the sense of *augmented* intelligence.

Augmented intelligence[1] is "a subsection of AI machine learning developed to enhance human intelligence rather than operate independently of or outright replace it. It is designed to do so by improving human decision-making and, by extension, actions taken in response to improved decisions."

The difference between artificial intelligence and augmented intelligence may be considered both a matter of degree and a matter of responsibility, as explained in [4, p. 191]: "The ultimate objective of augmented intelligence is using machines to enhance the work, expertise, and experience of a human user or worker." Augmented intelligence is therefore considered a *tool* that is no different from practical tools people have used throughout history to improve their "intelligence,"

---

[1]https://digitalreality.ieee.org/publications/what-is-augmented-intelligence#:~:text=Augmented%20intelligence%20is%20a%20subsection,in%20response%20to%20improved%20decisions.

from the printed book to the abacus. However, the difference between artificial intelligence and augmented intelligence may be a matter of degree. Augmented intelligence may transcend a limit line and function as a substitute for the human expert. Here, we encounter the issue of responsibility. Technological tools, such as GPT, may be considered augmented intelligence as long as they empower humans who take *responsibility* for using this tool. The contextual analysis presented in the current manuscript may cross a tipping point of expertise if it becomes fully automated. However, the responsibility remains within the domain of human beings.

As a young child, I was fascinated by Asimov's "I, Robot" [5]. The promise of AI and its necessary constraints echo in my mind almost 50 years on. This is why I prefer to approach the challenge addressed by this book in terms of augmented intelligence. When conceptualizing AI as a tool, the responsibility for its use remains entirely a matter for the humans who develop and use the technology.

In this brief manuscript, I have pointed out the importance of context in understanding human behavior because it is amazing to realize how context is ignored in a wide range of fields, from the study of personality to the automatic identification of social norms and their violations. One possible reason for this ignorance is a simplicity-biased engineering approach, focusing on simple solutions. But another explanation might just be the lack of tools to understand how to represent and use context for various tasks. I have tried to address this challenge by providing a deep theoretical understanding of context, proposing several approaches that can give us "operational" definitions of context, and using worked examples that show how to apply AI-based contextual analysis to get a better understanding of human behavior, from conversations going awry to reading great literature. AI is a promising tool for understanding context, specifically in domains that traditionally struggle with interpretation and context. LLMs are highly relevant for these fields (e.g., [2, 6]), which seem to be slow to adopt the new technologies. Therefore, the book is only a first step toward AI-based contextual analysis.

It goes without saying that it is impossible to cover all aspects of computational context analysis in a short manuscript. Therefore, this book should be read primarily as a first introduction to AI-based context analysis. Moreover, LLMs are a hot topic, with new technologies popping up almost daily. See, for example, the new LangChain[2] platform we used in some of our studies. This short manuscript presents an approach that these emerging technologies can significantly enrich. However, I have also emphasized the idea of augmented intelligence and the role of human researchers in the process. To repeat the metaphor used previously, the technology is the piano, and the researcher is the musician. Therefore, I conclude by inviting researchers to the challenging new domain of AI-based contextual analysis and urge them to empower their music through new methodological and technological tools.

---

[2]https://www.langchain.com/.

References

1. Neuman, Y., Cohen, Y., Assaf, D., Danesi, M.: Identifying the meta-forms of situations: a case study in computational semiotics. Int. J. Semiot. Vis. Rhetoric **1**(1), 56–71 (2017)
2. Neuman, Y., Cohen, Y., Yin, W.: Identifying social norm violation in movie plots: from Borat to American Pie. Digit. Scholarsh. Humanit. **38**(4), 1636–1645 (2023)
3. Neuman, Y., Cohen, Y.: AI for identifying social norm violations. Sci. Rep. **13**(1), 8103 (2023)
4. Sadiku, M.N.O., Musa, S.M.: A Primer on Multiple Intelligences. Springer, Cham (2021)
5. Asimov, I.: I, Robot. The Robot Series. Random House, New York (2004)
6. Neuman, Y., Danesi, M., Vilenchik, D.: Using AI for Dialoguing with Texts: From Psychology to Cinema and Literature. Taylor & Francis, UK (2022)

Author Index

© The Editor(s) (if applicable) and The Author(s), under exclusive license to   115
Springer Nature Switzerland AG 2024
Y. Neuman, *AI for Understanding Context*, SpringerBriefs in Computer Science,
https://doi.org/10.1007/978-3-031-64210-4

Subject Index